CAPITAL OF HAPPINESS

LORD GREY OF FALLODON AND THE CHARM OF BIRDS

CAPITAL OF HAPPINESS

LORD GREY OF FALLODON AND THE CHARM OF BIRDS

A Collection of Lord Grey's Writing
Selected and Introduced by
Jan Karpinski

Woodcuts by Robert Gibbins

Preface by The Rt Hon the Lord Home of the Hirsel KT DL

MICHAEL JOSEPH

To Melanie

First published in Great Britain by Michael Joseph Ltd. 1984
44 Bedford Square, London WC1.

Designed and produced by
BELLEW PUBLISHING COMPANY LTD.
7 Southampton Place, London WC1A 2DR.

British Library Cataloguing in Publication Data

Grey, Sir Edward
Capital of happiness.
1. Birds — Great Britain
I. Title II. Karpinski, Jan
598.2941 QL690.G7

ISBN 0-7181-2492-8

Printed in Great Britain

CONTENTS

PREFACE

by The Rt Hon the Lord Home of the Hirsel KT DL

Edward Grey was essentially a countryman, who, in his early twenties, found himself active in national politics, and that meant life in London.

He had to find a country refuge in which to refresh his body and mind, and that he did with regular visits to a cottage on the river Itchen, where he and his wife could indulge their skills in dry-fly fishing for trout, and their delight in the charm of birds.

The secret of success in trout-fishing and in bird-watching is to know when to be still; then all nature will come to be your companion. In his later years, when Grey lost his sight he would sit and wait for the birds to come to him.

I have two vivid recollections of him at Fallodon.

In the first he was sitting still and quiet, coaxing a flight of wigeon, wild and fresh from the North Sea, to feed out of his hand. They were doing so quite contentedly in a matter of days.

In the second he was in bed, a few weeks before he died, and the red squirrels were coming down the chimney covered with soot, which they spread all over the sheets as he shared his luncheon with them.

There is a delightful picture in this book of a walk with Theodore Roosevelt, in which Grey was the interpreter of the songs of English birds. All the cares of the world fell from their shoulders, while nature and beauty took charge, and so they will for the busy reader who is lucky enough to follow this story of an attractive and talented man.

Home

INTRODUCTION

Nearly sixty years ago, Sir Edward Grey, Viscount Grey of Fallodon, then almost blind and nearing the end of a life which had brought him both public distinction and personal grief, dictated a book called *The Charm of Birds* to his wife. As Grey's first biographer, G.M. Trevelyan said, it was as if Grey's memories of bird life were a 'capital of happiness' on which he drew as blindness and sorrow befell him.

The Charm of Birds was an immediate and continuing success. The popularity which it enjoyed in its day did much to promote both bird-watching as a hobby and the conservation of wildlife as a cause, at a time when little grebes nested in St James's Park in the heart of London and red-backed shrikes could still be seen in Hampshire. The main purpose of this book is to make Grey's writings on birds accessible once more by assembling an anthology of his best work, drawn mainly from *The Charm of Birds* itself. The factual background necessary to place Grey's writings in their historical and personal context is provided in the first part of the book: Part II is devoted entirely to extracts from Grey's books, illustrated with the original woodcuts by Robert Gibbings.

Grey was a reticent man, with a highly-developed idea of what it was proper for the outside world to know about his private life. There is still a suggestion of mystery about him which is deepened by the disappearance of the bulk of his private papers after his death. His qualities and achievements are, however, undoubtedly deserving of reappraisal and his 'capital of happiness' provides us with considerable insight into the life and personality of a remarkable man.

PART I

CHAPTER I

GREY'S LIFE

Born plain Edward Grey in 1862 he died in 1933, seventy-one years later, as the Right Honourable Sir Edward Grey, Baronet, Knight of the Garter, Viscount Grey of Fallodon. Widely remembered as the Foreign Secretary who took Britain into the First World War, it was Grey who uttered the famous words 'The lamps are going out all over Europe; we shall not see them lit again in our lifetime' shortly after dispatching the ultimatum to Germany on 4 August 1914. In startling contrast, Grey wrote *The Charm of Birds,* one of the most successful nature books of its day, which went through ten editions in the decade after it was published in 1927. These two facts are remembered though much else is forgotten, and for this there is a simple explanation.

Without a doubt the Great War (as contemporaries called it) set the shape of our modern world. Grey's part in its outbreak was critical. His interest in birds is a useful human detail with which to season the otherwise bland account of alliances and *ententes* (it is one of those happy facts of history which can be used more or less to taste) as evidence of the classbound narrow-mindedness of a country squire who preferred animals to people or, alternatively, of the wide-ranging compassion of a truely heroic world statesman. From either viewpoint it was no more than an amiable hobby.

There is nonetheless something intriguing about Grey's bird-watching. Why did he write so well and so successfully about it? Why did he take it up in the first place rather than any of the other pastimes open to an Edwardian gentleman? It is somehow difficult to imagine *The Fun of Golf,* say, or *The Joy of Stamp Collecting* going through ten editions in so many years. The fact is that if we want to understand Grey as a personality, rather than as a mere component of the vast, bureaucratic engine of European diplomacy, then we have to explain his love for birds. Conversely, Grey the nature writer cannot be isolated from Grey the politician and cabinet minister. He wrote for a generation to whom the outlines of his life and career were familiar — they must therefore be familiar to us too. We must know something of the world in which he lived, of the accidents of circumstance and opportunity which determined his role in it.

The first accident is birth. In Grey's case it entitled him to participate in that most exclusive game of chance, the government of the United Kingdom. The question of his early years was whether he would play the cards which heredity had dealt him; thereafter, of how long he would stay in the game and with what success. It is impossible to understand Grey's life without some explanation of his ancestry.

He was descended from an old Northumberland family which had first distinguished itself in the French wars of the 15th century. This accounted, no doubt, for the family motto *'De bon vouloir servir le Roy'* ('To serve the king willingly') and the family crest, a scaling ladder (for climbing the walls of besieged castles). This martial tradition persisted. During the American War of Independence Grey's great-great-grandfather won fame as 'No-flint Grey' because he once ordered his troops to knock the flints out of their muskets to ensure surprise and success with the bayonet. In due course he was elevated to the peerage as the first Earl Grey. The first Earl's sons and grandsons included an Admiral, a Captain (RN), two Generals, a bishop and four MPs, of whom one became Prime Minister, one Secretary of State for the Colonies and a third became Home Secretary. Grey was descended from one of the first Earl's younger sons, the Captain (RN). This junior branch of the family was known as the 'Greys of Fallodon' since Grey's grandfather (Sir George Grey, the Home Secretary) had taken up residence at Fallodon Hall in Northumberland, a few miles from the more splendid accommodation of the senior branch (that of successive Earls Grey) at Howick.

Sir George's only child, a son known in the family as 'young George', showed little inclination to expand the catalogue of family achievement. He joined the army and saw some active service in the Crimea and India, but resigned his commission at the age of twenty-seven to devote himself to his duties as equerry to the Prince of Wales, to the management of the Home Farm at Fallodon and most especially to raising his seven children, the eldest of whom was the future Foreign Secretary and bird-watcher. The whole family lived happily together with Sir George and Lady Grey at Fallodon. Grey later wrote that:

> My Mother was one of the gentlest of human beings. She shrank even from argument and in order to make life go smoothly was always prepared to efface

> herself. This made it possible for the arrangement for a joint home between my Grand-parents and Parents to work smoothly, and my early recollections are of an uneventful and extremely happy home. What we first know in childhood we imagine to be typical and the usual way of the world, and I was brought up assuming that the world was a happy place, in which no grown-up people did anything that was wrong.

The first of many severe shocks to Grey's view of the world as a happy place was the death of his father after a sudden illness. Grey was twelve years old and this was a great blow, with important consequences for the future. It meant that he was likely to inherit the family title and estate at a comparatively early age. A time-consuming and expensive career in politics would thus be open to him. MPs were not paid until 1911 and if his father had lived, Grey would have had to find some respectable livelihood like his younger brothers, two of whom went into the colonial civil service while the third went into the church.

Sir George sent his grandson to Winchester and then to Balliol College, Oxford. At both places he was identified as being able but idle. At Winchester he developed an intense and enduring passion for dry fly fishing on the River Itchen, if little else. When Grey went up to Balliol in 1880 its famous master, Benjamin Jowett was then at the height of his powers. His aim was to groom able young men for distinguished careers in public life. Asquith, Grey's future ally and Prime Minister, the Marquess of Lansdowne who preceded Grey at the Foreign Office and George Nathaniel Curzon who came after him — all were Balliol men. Several of the ambassadors who served under Grey had been his contemporaries at Balliol, including those in the key posts at

Washington and Constantinople in 1914. Jowett had once pointed out Lansdowne as a man who could not help becoming Foreign Secretary: he had no cause to say the same about Grey who, like many another undergraduate, was generally self-indulgent, sometimes drunk and occasionally disorderly. He himself later described this more pleasantly as clearing the way for more serious things. His eventual nemesis was recorded by Jowett in the Balliol minute book in the following terms:

> Sir Edward Grey, having been repeatedly admonished for idleness, and having shown himself entirely ignorant of the work set him in vacation as a condition of residence, was sent down, but allowed to come up to pass his examination in June.

One of the many attractive facts about Grey's career is that having left Oxford in disgrace with a third in Jurisprudence he lived to become its Chancellor in 1928.

One might have predicted that, after his Oxford débacle, Grey would have retired to Fallodon (which he had recently inherited from his grandfather) to pass his time in agreeable indolence. Instead he developed a sudden interest in politics, serving the customary stint as an unpaid private secretary (through the good offices of his second cousin, Lord Northbrook, then First Lord of the Admiralty) and making himself known in Northumberland by speaking at public meetings. He made a good impression on both fronts and the notion was soon abroad that Grey might be the man to recover the constituency of North Northumberland from the Tories, to whom his grandfather had lost it in 1852. There were, however, other likely candidates, and Grey's views were a trifle too advanced for some. Nevertheless after a wrangle, in which Sir George's old friend Sir Charles Trevelyan played a decisive part, the nomination went to Grey. Another of Sir George's old friends, Mandell Creighton, vicar of Embleton (in which parish Fallodon is situated) and later Bishop of London, became Grey's confidant and adviser.

Meanwhile, he had fallen in love with a girl called Dorothy Widdrington. Like him she came from a well-to-do Northumberland family. She was good-looking, high-minded and intelligent, but withdrawn, shy and not a little neurotic. She was perhaps not the ideal choice for a budding politician, but Grey was guided by his affections rather than his ambitions. The couple were married in October 1884 in the middle of the election campaign. It was remarked that the new Lady Grey seemed to have enjoyed the contest, as well she might, considering the bouquets, the cheers and the compliments which a candidate's attractive young wife could expect in those days. However, she always regarded Grey's involvement in politics as an experiment, which could be terminated if the results were unsatisfactory.

Grey entered Parliament at an exceedingly delicate stage. The Liberals were divided on a whole range of issues, but most particularly on Ireland. Grey favoured Gladstone's policy of Home Rule. Many of his erstwhile supporters did not and in the next election they fielded a 'Liberal Unionist' candidate against him, substantially reducing his majority. During this difficult period Grey laid the foundations of his career, steering a careful course, championing the vaguely progressive, eschewing the downright revolutionary, tempering idealism with expediency. He forged an alliance with two of the up-and-coming men of his generation, Richard Haldane, later Secretary of State for War, and Herbert Asquith, later Prime Minister. In 1892 Grey had his reward: when Gladstone formed his last ministry he made Grey the Under Secretary for Foreign Affairs under Lord Roseberry, who seemed set fair to succeed to the Premiership in due course.

Grey had no particular qualification for his new post, but nevertheless did well. His task was mainly to present the Government's policy to the Commons, which might not be easy considering that he played little part in its formulation and there might well be considerable differences of opinion in the cabinet. It turned out to be an invaluable apprenticeship. He learnt from personal experience how unsatisfactory and dangerous was the traditional concept of splendid isolation. It laid Britain open to German blackmail over Egypt and to the risk of war with France over some trivial misunderstanding. Most important, he became known to a wider public through the 'Grey Declaration'. This was an almost impromptu warning to France that any incursion into the Sudan would be regarded as an unfriendly act. It assumed some importance subsequently when a secret French expedition appeared at Fashoda on the Nile in 1898.

In 1894 Gladstone at last retired and Roseberry enjoyed a short but disastrous career as Prime Minister. His government fell in 1895 and the Liberals lost badly at the ensuing General Election. It was ten years before they returned to power. The early years of opposition perhaps mark the apogee of Grey's early life. He had a bright future. He was extremely fit and energetic (as indeed he remained all his life), a prodigious cyclist and walker and, in 1896 the undisputed real tennis champion (there were in fact two championships, one at Queen's and one at Lord's — in 1896 Grey won both). In 1890 he had built a little wooden cottage on the banks of the river Itchen, which enabled him to put in a fair amount of fishing, to the extent that in 1898 he wrote a book on the subject, called simply *Fly Fishing*. The Cottage was much more than a fishing lodge. It played a central part in his private life, the intimate, slightly reclusive world of nature and of birds which he had created with Dorothy. Od-

dly enough, this private world had a public aspect to it as well. Grey had befriended W.H. Hudson, who was the leading nature writer of the day. Two chapters of one of his best-known books, *Hampshire Days* published in 1903, were based on a visit to the cottage, while the book as a whole was dedicated to the Greys — 'Northumbrians with Hampshire written in their hearts'. Grey was later to recall that,

> ...though I could not know this at the time, the best of my trout fishing was already over when *Fly Fishing* was being written...

It was typical of Grey that the greatest threat to his private happiness came from the prospect of public success. Throughout his marriage to Dorothy there was a continuous tension between his advancing political career and her disenchantment with life in London, which to a large extent he shared.

Two important issues had to be settled before the Liberals

could return to power. The first was the Boer War, over which the party was split. The majority opposed the war. However a section headed by Haldane, Asquith and Grey supported it as just and necessary, while attacking the Government's handling of it. This split could only be healed after the war came to an end in 1902. The second issue was the leadership of the party. This was greatly complicated for Grey by the perverse isolation of Roseberry to whom, as far as he reasonably could, he remained loyal. As Roseberry progressively faded out of the picture, Grey became increasingly convinced that Asquith was the only man who could successfully lead the party in the Commons. This conclusion, as Grey later admitted freely, was based on an underestimate of the capabilities of Sir Henry Campbell-Bannerman, who was favoured by the rest of the party. Grey, Haldane and Asquith plotted unsuccessfully against Campbell-Bannerman but he included all three of them in his government when he went into office in December 1905. Grey became Foreign Secretary, a post he was to hold for eleven years.,

As far as history is concerned, one issue dominates Grey's tenure at the Foreign Office, and this is the outbreak of the First World War in 1914. No account of Grey can really dodge this issue and it is dealt with more fully in Chapter III.

These were also momentous years in English political life. Grey was not so immersed in the work of his department that this escaped his attention. Whereas in foreign affairs he was content to develop the logic of the *Entente Cordiale* between Britain and France which had been established by Lansdowne, in domestic matters he tended to take a fairly progressive stand. For example, he played an active role during the great constitutional crisis of 1910 pressing strongly for an elective second chamber — a suggestion which still seems fairly radical today. He also spoke up in favour of votes for women and supported Lloyd George with his important welfare legislation.

Sadly, Grey had many personal matters on his mind too. He had only been in office for a matter of weeks and was in the middle of a dangerous international crisis, when Dorothy was fatally injured in an accident in Northumberland. This was a terrible blow, he missed her bitterly for the rest of his life. Then, Grey began to go blind. He had taken up squash, since it was difficult to find the opportunity to play tennis any longer, and started to have trouble seeing the ball. By the time he left the Foreign Office in 1916 his sight was virtually ruined. he could no longer see to fish, to play squash or tennis, or to watch birds except at the very closest quarters. His marvellous ear for their song, however, remained.

Grey left office in 1916 as a result of the upheaval by which Asquith was replaced by Lloyd George, cast as a sort of war supremo. Grey was only fifty-four — young enough to have continued an active political career. Instead he chose to play the part of an elder statesman from the sidelines. As such, disaster and distinction descended on him in about equal measure. The first instance of the latter was his elevation to the House of Lords as Viscount Grey of Fallodon. This had taken place while he was still Foreign Secretary with the intention that his duties would be less exhausting to perform from the Upper House. Besides this, he became a KG, Chancellor of Oxford University and an honorary DCL, and the Vice President of the National Trust.

The last half of Grey's adult life was punctuated by a series of personal disasters. His three younger brothers all died before he did. Ironically and tragically, the two of whom he was fondest both achieved the ultimate apotheosis of the 'Boy's Own Paper' hero: they were killed while big game hunting in Africa; one by a lion in 1913; the other by a buffalo in 1928. In 1917 Fallodon was burnt to the ground. He wrote sadly to a friend:

> I kept everything in the house in the place where Dorothy put it and now I shall never see the familiar rooms again. And of course my nursery and the room where I learnt the alphabet and memories like that are swept away. It makes me feel that affections should be set on the open air and the seasons and the stars and things that are less perishable. But after all how gladly would I have had the house burnt if only Dorothy might have lived.

Fallodon was to be re-built. However, in 1923 the Cottage too was burnt down. The distressed caretaker announced this blow in a letter that was succinct and to the point:

> My Lord. What can I say to you? I am sore vexed but your pretty cottage is burnt to the ground. What will you say my Lord?

There was nothing to be said. The Cottage was left as a ruin and thus it remains to this day. G.M. Trevelyan wrote in his biography that:

> ....it became part of Edward Grey's heavy fate that every person with whom he desired to make a home for life was doomed to die, and that every habitation in which he loved to dwell was burnt to the ground.

The one happy event in Grey's later life was undoubtedly his marriage to Pamela Tennant, widow of Lord Glenconner, in 1922. They had been close friends for many years and had both suffered bereavement — Lady Glenconner had lost a favourite son in the war, as well as her husband in 1920. It was largely owing to her interest that Grey started to write and give talks. His first book *Twenty-Five Years* (1925) gives an account of his work at the Foreign Office, the *Fallodon Papers* (1926) is a collection of talks and articles, and finally in *The Charm of Birds* (1927) he set out some of his vivid memories of bird life. Nor did Grey retire altogether from public life. He led the Liberals in the Lords for a time, thus earning the unrelenting hatred of Lloyd George with whom he clashed. He was an ardent supporter of the League of Nations and, appropriately, his last debate in the House of Lords was about the pollution of the sea by oil spillage.

Grey died after a short illness at Fallodon Hall on 7 September 1933. His obituary filled a whole page of *The Times* . A crowd of dignitaries packed Westminster Abbey for his memorial service. But his ashes were buried with characteristic modesty beside those of Dorothy in the grounds of Fallodon. The inscription on their grave states simply,

> HERE AMONG TREES THAT THEY PLANTED TOGETHER ARE PLACED THE ASHES OF EDWARD AND DOROTHY GREY.

CHAPTER II

GREY AND BIRDS

There were two facets to Grey's love of birds: bird-song and ducks. Both arose in the spring of 1884, shortly after he was sent down from Oxford. In *The Charm of Birds* he described the incident which, as he subsequently explained to a friend, first made him notice the singing of birds: ' ... a loud song proceeding from a thick cypress arrested my attention ... I wondered what bird this was, and struck the branches with my stick. Out flew an unabashed wren ... ' At about the same time he was starting his famous collection of waterfowl at Fallodon and by the spring of 1885, he had assembled seventeen different species.

What immediately inspired these interests it is difficult to say. Grey was a young man of twenty-two discovering for the first time, in the way that young people do, a great many things which were to be important in his future life; poetry, music, literature, but above all, his political vocation and his future wife, Dorothy.

It was undoubtedly due to Dorothy that birds became such a central part of Grey's world. This had something to do with Grey's other great interest, angling. Dorothy enjoyed fishing, but perhaps not as much as her husband. She would take a book to read in odd moments and sometimes she would just sit and watch. We may guess that, as an alternative to boredom on the river banks she felt the need for some other interest to share on their fishing trips. The little song-birds which abounded in the trees and bushes at the Cottage and in the grounds of Fallodon offered an obvious possibility. At any rate, they soon became an important bond between Grey and Dorothy.

The strength of this bond was undoubtedly increased by the fact that the Greys never had children. Of course, children would have made fishing at the Cottage almost impossible and would have absorbed a great deal of the care and attention which Grey put into his ducks. But the couple's

childlessness had a deeper significance perhaps. It was widely believed that the Greys' was 'un marriage blanc' to use the quaint euphemism of the day. Indeed, Grey once confided in a friend that after their honeymoon Dorothy had developed an aversion to 'the physical side of marriage' and that they had agreed to live together 'as brother and sister'. Nowadays we would think such a platonic relationship between a young and healthy couple as extremely odd, not to say distasteful. Dorothy would be urged to consult a psychotherapist and her husband to commence divorce proceedings, alleging her 'unreasonable behaviour', lest he be suspected of latent homosexuality. We are perhaps less tolerant than we like to suppose: Grey's colleague Lloyd George carried on with impunity the sort of infidelity for which Mr Cecil Parkinson was hounded from office. The great advantage of Victorian morality was that provided one complied with the formalities, a considerable latitude was allowed as to the substance. In particular, a married couple were free to define the terms of their own relationship without having to conform to a stereotype of physical passion.

Sex is not only a part of the procreative mechanism. It is also a way in which two people can demonstrate their love for each other. There can be no doubt that the Greys were very much in love and that when they watched birds together or spoke or wrote to each other of birds, this was one way of showing it. The lack of any physical expression of their love would account for the particular intensity of their mutual enthusiasm for bird life.

There was another personal and deep-rooted factor which increased this enthusiasm. Grey was what we would now think of as a very religious person, though not in any bigoted or narrow-minded sort of way. It was part of his faith that 'God sees all things.' He once wrote of spring at the Cottage:

> The beauty has been overwhelming; pear and apple blossom overlapped and the profusion and splendour were more than human capacity could appreciate. I used to feel at this season of the year a sense of waste because I could not enjoy at once all that was spread

> abroad; till one day the overwhelming egotism of looking at it from this point of view occurred to me, and I thought that God might be contemplating it all. Then I ceased to be oppressed by the sense of waste ... The sight of all this beauty and the feeling of response to it in oneself gives assurance that God rules in the Universe, and that evil cannot prevail.

Grey believed that the object of life was to try to make oneself a better person. The most direct way of doing this was by contemplating God's works, that was to say, nature. Thus Grey's love of birds was an expression not only of his love for Dorothy, but also of his love for God.

After Dorothy's tragic death, birds became important to Grey in a different way. In the words of G.M. Trevelyan, 'After 1906, his powerful memory enabled him to live on the capital of happiness he had already amassed, and to watch the birds beside the Itchen as he walked the London streets.'

Birds were important too in the relationship between Grey and his second wife, Pamela. A certain reticence is detectable in what Trevelyan wrote about Grey's second marriage, which may reflect some unease which Grey himself felt. There is bound to be a certain awkwardness in any second marriage, particularly between a widow and a widower who both believe in an after-life, for whom the taking of a new partner must suggest some breach of faith with the memory of the old one. There will be painful reminders of the past and in order to avoid these, much of the usual small talk may have to be eschewed. For years before their marriage, Grey and Pamela had carried on a lengthy and, one suspects, rather literary correspondence '... concerned almost entirely with verse, and with bird lore'. Grey had a great partiality for Wordsworth, while Pamela was an established poet in her own right, birds being one of her main sources of inspiration. The subject of birds formed a neutral zone which Pamela could quite properly enter with Grey during the lifetime of her first husband and which did not evoke any painful thoughts after his death.

It was due to Pamela that Grey started writing for publication. In the dedication to his first book *Twenty-Five Years,* he said that without his wife's '... constant help and encouragement the work would never have been done'. The same was true of *The Charm of Birds*. Indeed, quite probably he dictated the book to her and quite possibly he did so mainly at her home, Wilsford Manor. Although *Twenty-Five Years* was a greater achievement from an intellectual point of view it was *The Charm of Birds* of which Grey was particularly proud. When he spoke of 'my book' it was *The Charm of Birds* to which he was referring.

Lloyd George is said to have accused Grey, in private and behind his back, of some scandal in his private life. If this was so, there is no exact record of the allegation, while we have *The Charm of Birds* as proof that Grey successfully resolved the tensions arising from the unhappinesses of his life and also the circumstances of his two marriages. Grey's writing does not contain the self-revelation we might expect from an embittered or unbalanced personality. Nevertheless, it does tell us some interesting things about him, and, of course, it reveals a great deal about his attitude to birds.

First of all there is his great interest in bird-song. This is the central theme of the first half of the book, extracts from which are given in Chapters VI to IX. Grey did not have much interest in seabirds or birds of prey, whose utterances are not 'song' to the ornithological ear, and had little to say about them. He candidly acknowledged this and added that 'the omissions occur because the purpose of the book is to deal only with those aspects of bird life that have happened to be a source of interest and pleasure to the author ...' However, Grey did not abide strictly by the narrow definition of the term 'song', for '... it would impoverish pleasure in birds so to restrict the thought of song. Every bird should be included that has what appear to be joy-notes associated with courtship or some pleasant emotion, entirely distinct from the call or alarm notes.' Did the acuity of Grey's ear for song owe something to the increasing weakness of his eyes? On reflection, one would doubt this. In order to learn to identify the songs of different species it is necessary both to hear the song and to see the bird which sang it, or at least this was the case before sound recording. Grey confirms this when, in a discussion of the different songs of the various species of tit, he says 'The willow-tit I have not yet distinguished from the marsh-tit, and with impaired sight I could not now discover it for myself.' There are several slightly surprising omissions from Grey's writings apart from seabirds and birds of prey. In *The Charm of Birds* for example, he mentions starlings, rooks, crows and jackdaws, but not jays, magpies or ravens. The simple explanation must be that he felt he had nothing interesting to say about these birds. Not only was Grey no name dropper when it came to birds, he was also not what we would today call a 'twitcher'. He did not set out with the object of spotting as many different species as possible. Very few of the species to which he refers could be described as rare. Of those described in the extracts below, perhaps only the cirl bunting and the great grey shrike (both in Chapter IX) fall into this category. Both were more common sixty years ago and in any event the cirl bunting happened to nest in the ivy growing on the Cottage, while a great grey shrike flew more or less under his nose one day. For all this Grey displays a remarkably wide knowledge of British bird life, particularly bearing in mind that both our information about birds and our skills in observing them have vastly improved since his time. For example, W. H. Hudson, who was supposed to be something of an authority on birds, could only speak of some hundred species from personal knowledge and had difficulty in distinguishing a lesser black-headed gull from a herring gull in the winter. In *The Charm of Birds* Grey mentions 125 British species from personal knowledge and could certainly have spoken of many more. He could also distinguish a lesser from a common whitethroat with apparent ease.

As for Grey's interest in waterfowl, described in the extracts in Chapters XI and XII, there is surely something instructive in his decision generally not to pinion his ducks, that is to amputate the upper part of one wing so that they could no longer fly away. 'With pinioned birds there is a doubt as to the reason for their complacence: with free birds their tameness is a compliment.' Grey's pleasure was not just to see his ducks swimming about his ponds, but '... to see a pair of Chiloe wigeon on the wing high in the air, conversing audibly as they fly, and to know that when they alight they will take food from the hand ...' There is something telling too in the almost parental concern for the fate of his favourite individual ducks, such as Elizabeth the pintail.

Grey was curious about many of the unanswered questions of bird behaviour and was quite prepared to put forward his own speculative suggestions. He was, however, inclined to deprecate scientific investigation. At the end of a long discussion on why birds sing he wrote, 'Perhaps the conclusion of the whole matter is that we should enjoy the song of birds without questioning too closely why they sing. Enjoyment may be impaired by teasing oneself with trying to understand.' In fact Grey's speculations on bird behaviour are perhaps the least effective part of his writing, while his enjoyment comes across fresh and strong. This is partly because he adopts the same approach to ornithology as he did to foreign policy: questions were to be answered, if at all, by a process of logical reasoning applied to the established

facts. The idea of devising experiments to establish new facts would have been no more acceptable than a foreign policy adopted for no better reason than to see how other nations would react to it. It was one thing to seek to share God's Truth; another to pry into His secrets.

It would be wrong to suggest that Grey completely rejected all general theories of nature. Indeed he proposed one of his own, '... that there is in Nature some tendency or purpose ever at work for the sake of creating infinite variety'. However, this idea that there was some random factor at work in nature, was based on a dissatisfaction with what he referred to as '... theories of evolution, survival of the fittest, natural selection and so forth'. Grey said of such theories that:

> In birds, for instance, it is very satisfactory to find and note some habit, or some protective colouring of plumage or eggs that has been evolved and to which the survival and welfare of the species is due. Yet it seems to be beyond the utmost ingenuity of human intellect to stretch any of these natural laws or processes sufficiently to make them cover and account for all that we see.

With hindsight we can see that Grey underestimated not only the ingenuity of the human intellect and the elasticity of the so-called natural laws, but also the amount which could be learnt from methodical research and careful observation.

One very important point emerges from this. There was a considerable vogue in Grey's time for what is now known as 'Social Darwinism'. This was essentially a distortion of Darwin's theory of evolution, an attempt to apply it to contemporary social and political situations. Thus New York millionaires argued that starvation was nature's way of eliminating poverty, so that it was wrong to give food or money to the starving poor. Likewise German generals held that the military conquest of weaker nations by their stronger neighbours was all part of nature's plan, so that any policy which aimed at preserving peace was in the long run both futile and undesirable. We know that many people in Europe were influenced by nonsense of this sort, but to what extent did it influence Grey? The answer is 'not one jot'.

Social Darwinism encouraged many of Grey's contemporaries to look on war as a remedy for materialism. This point of view was a bit like the public school attitude to caning: it was painful but, if taken like a man, good for the character. Grey too became disenchanted with the Victorian concept of progress, and the materialist values which went with it. However he had a rather different answer which depended on an interest in nature and did not involve death or maiming.

Although it was mainly through his writing that Grey became renowned as a bird lover, he did in fact publicly identify himself with the cause of bird protection from his early days in politics. He was one of the earliest members of the Royal Society for the Protection of Birds, which he joined in 1893, becoming a Vice President in 1895. In 1919 he was asked by the Government to chair a committee which was to review the complicated and cumbersome legislation of the time, the Wild Birds Protection Acts 1880 to 1896. The result of the committee's labours was a bill which would have greatly simplified the law, made it easier to enforce and for the first time provided a proper legal framework for bird sanctuaries. Grey piloted this bill through the House of Lords in 1923, but the Government failed to find time for it in the Commons and it did not reach the statute book. It did, however, exercise some influence on future legislative thinking, although it was not till 1956 that Parliament finally got round to the job.

Grey's most effective contribution to the cause of bird protection was really as a writer. It was undoubtedly enhanced by his political achievements and social position. The English country gentleman has always enjoyed nature, but his enjoyment tends to be keenest when viewing nature down the barrel of a 12 bore. This was not so with Grey. However, he was no prig. If his neighbours refrained from shooting one of his rare but unpinioned ducks, it was an act of kindness on their part: it was not in him to object to their shooting wild duck at all. This sense of tolerance makes Grey a rather more effective propagandist than many of today's banner-waving campaigners.

CHAPTER III

GREY AND THE FIRST WORLD WAR

Grey was Foreign Secretary for eleven years to the day: a longer period of office than any other Foreign Secretary before or since. This was to influence his subsequent writing about birds in two important ways. First of all, it meant that for a long time he had very little leisure to spare for Fallodon or the Cottage. He described a typical day in his life as Foreign Secretary in *Twenty-Five Years*. He arrived in the Office at 11 a.m. and spent the morning reading telegrams from British representatives abroad. Between 1.30 p.m. and 3 p.m. he went out to lunch. When he came back from lunch he would talk to foreign ambassadors or other important people who wished to see him. He would then do paperwork until 6 p.m. or 7 p.m., when he would go home for dinner. After dinner he would continue with paperwork until bedtime, leaving a pile of papers beside his bed to work on for an hour or so before breakfast at 9 a.m. After breakfast he set aside an hour for his personal correspondence before leaving for the Office to start the gruelling daily round again. At any rate this was the routine when there was no great crisis, and he was not required at a cabinet meeting or in the House of Commons. If any of these things happened there would be a backlog of work to make good somehow. It was vital not to allow any arrears to accumulate: every day of the year a flood of telegrams would pour in from all corners of the globe raising questions to be investigated, pondered and decided. Thus holidays had to be short, and were always likely to be interrupted by some crisis, while bundles of Foreign Office paperwork always followed him down to the Cottage or the New Forest at weekends.

Although *The Charm of Birds* was not written until about ten years after Grey had left the Foreign Office, inevitably this important period in his life influenced the way in which he wrote. There is something of a slide show about Grey's style: a bright and colourful picture appears before us; an unseen voice talks about it for a minute or two; then another picture, and the voice continues. Each picture is part of a carefully assembled, carefully maintained collection. One can imagine Grey, in the occasional moment of leisure during the eleven years of loneliness and strain, picking out one of his treasured memories, rehearsing it in his mind, refreshing and restoring it, before filing it away again for future reference. One cannot help feeling that had he never been Foreign Secretary, Grey would not have written so well about birds.

Certainly, if Grey had not been Foreign Secretary *The Charm of Birds* would not have sold so well. A book by a former minister of the Crown is a very different marketing proposition to one by a dilettante country squire. However, the interest of the book, at the time it was published, went far beyond finding out what a successful politician did in his spare time. To explain this we need to consider the origins of that vast human catastrophe known as the First World War.

When Grey became Foreign Secretary a major European war was no more than a possibility and, moreover, it was only one of a number of possibilities with which he had to reckon. Britain's extensive interests around the globe involved a number of different risks of conflict with other powers, both European and non-European; for example with Russia over Persia or with the USA over Central America. What then was Grey's policy? Essentially it rested on two simple propositions, though Grey himself never put it in this way. The first proposition was that Britain's interests, especially her commercial interests, depended on peace and stability in the world. In 1905 most Liberals would have agreed with this wholeheartedly. But behind it lurked a second proposition, effectively a proviso to the first, that in certain circumstances British security might require her involvement in a continental war. After two world wars and thirty years of NATO this seems sound, not to say trite, but the bulk of Grey's party would have rejected it as dangerous madness. Nonetheless, it guided certain key aspects of his policy, from his first days as Foreign Secretary. There was no logical inconsistency between the two propositions. However, one would have commanded general support, the other would not. There was therefore an element of political schizophrenia about Grey's policy.

Germany played the leading part in the affairs of Europe prior to 1914. She had a common frontier with each of the three other 'Great Powers' of the European continent, France, Austria-Hungary and Russia, while she was the only country which could threaten Britain across her open frontier, the sea. Altogether, Germany's position was analagous to the banker in a game of pontoon.

Two words sum up German foreign policy leading up to 1914. The first was *Weltpolitik,* or world policy, which encapsulated the idea that Germany's European pre-eminence entitled her to play a more prestigious part on the world stage and to behave with increasing aggressiveness until this was allocated to her. The second was *Einkreisung,* or encirclement, the plot of Germany's jealous neighbours, egged on by Britain, to hem her in until a suitable opportunity occurred to fall upon and destroy her. These two ideas were regarded by many Germans as the irresistible force and immovable object of European affairs, which sooner or later were bound to meet in devastating collision. In fact they were cause and effect: the more aggressive the pursuit of *Weltpolitik* became, the more hostile were Britain, France and Russia to it.

Did the Germans deliberately plan to achieve the goals of *Weltpolitik* by military conquest? Certainly they had just such a plan, although its political results were of peripheral concern to the soldiers who formulated it. The rationale of the Schlieffen Plan, as it is known after the Chief of the General Staff originally responsible for it, was as follows. Germany was surrounded by implacable enemies. Only Austria-Hungary, a fellow Teutonic power, was friendly. However, neither France nor Russia were strong enough to defeat Germany on their own. They would be likely, therefore, to join together to impose a war on two fronts on Germany. However, they could do this in a purely defensive way, thus cheating Germany of the decisive victory which was essential to her success in the war. The Russian forces could retreat into the vast interior of their country; the French could sit tight behind their strongly fortified eastern frontier, against which frontal assault would achieve nothing.

But across France's northern frontier, the road from Brussels led straight and wide to Paris, bypassing all the French fortifications. This was therefore the way which the right wing of an attacking German army should follow, while the centre and left pinned down the French army in the east. To the German Generals who formulated this bold idea, political considerations had no part in the conduct of war. yet it was precisely the political consequences of the Schlieffen Plan which were its most remarkable feature. First, it involved a flagrant violation of Belgian neutrality, which Britain had promised to defend. Would the British go to war on behalf of the Belgians? The German generals weighed the contemptible little British army in the military balance and decided that the possibility of British intervention was of no importance. Second, there was the real possibility that when it came to the point, the French would not be prepared to go to war for the sake of Russia. However, the German generals were not to be thwarted by any reluctance on the part of the French to suffer the disaster which was intended for them. The Russian army had to be defeated by proxy on the Seine: the generals had no other plan. Third, there were the political results which would follow the success of the Schlieffen Plan. It is unlikely that the Germans gave this much thought before the war, but it was plain what must happen. Belgium would become a satellite of the German empire, France would cease to be a great power and Germany would be in a position to dictate to the rest of Europe, Britain included.

Such was the logic of German war plans from 1905-14. How did Russia, France and Britain react to it, or rather to what they suspected of it? The threat which *Weltpolitik* implied to Russia and France had been apparent for twenty years: in 1894 the two countries had entered into a Military Convention to come to each other's help in the event of an attack by Germany. The position of Britain was more problematic. Of one thing most Britons were certain: they wanted nothing to do with 'continental entanglements' which would involve them in other people's quarrels.

Grey, however, realized that a defeat of France by Germany would have unpleasant consequences for Britain which could only be effectively forestalled by timely military assistance where it was needed most, and that meant deploying a British Expeditionary Force in Northern France. Careful planning was necessary to achieve this, and during his first months in office Grey authorized discussions between the British and French staffs. Strictly speaking, Britain had only an 'understanding' with France, but it acquired all the trappings of a military grand alliance. In fact, the logic of the *Entente Cordiale* was as ruthless and cynical as that of the Schlieffen Plan. It required Liberal Britain and socialist France to side with repressive, autocratic Russia in a war which might easily start over some issue which was entirely peripheral to French or British interests.

The final crisis of the old Europe was precipitated by the assassination of the heir to the Austro-Hungarian empire, the Archduke Francis Ferdinand, by Serbian nationalists at Sarajevo on 28 June 1914. The Austro-Hungarian generals had long wanted to conquer Serbia and now the politicians allowed them their chance. They calculated that Serbia's powerful ally, Russia, would be neutralized either militarily or diplomatically by Germany. The Germans expected to win an easy diplomatic victory, but if this was not forthcoming, they were by no means averse to the alternative. When it became apparent that the Russians would not abandon Serbia, the slide into war became swift and irreversible. On 25 July the Austrian ambassador left Belgrade, the Serbian capital. Exactly a week later, on 1 August, the first German troops crossed the frontier into Luxemburg.

There can be no doubt that during this crisis Grey did his best to maintain peace. Afterwards he was to lie awake at nights agonizing over whether or not he could have done better. The question was entirely academic. None of the expedients which, with hindsight, his detractors put forward could have prevented war. The Austro-Hungarians had made up their minds that they would not be prevented from having a war against Serbia, the Russians that they must go to Serbia's aid, the Germans that war was bound to happen sooner or later and that this seemed as good a time for it as any. The threat of British intervention would not deter the Germans or Austro-Hungarians. Conversely, the threat of

non-intervention would not persuade the Russians or the French to abandon what they saw as their vital interests.

War brought Grey's political chickens home to roost. Initially the majority of the cabinet were determined not to get involved. With great skill, Grey and Asquith managed to avoid bringing matters to a head. Slowly the enormity of the German war plans began to sink in on the isolationist ministers. So did the popular enthusiasm with which the prospect of war was greeted. By 3 August, when Grey had to make a statement to the House of Commons the majority of the cabinet had come round to his way of thinking. His statement to the House clinched it and only two ministers resigned. The Government, the Liberal Party, Parliament and the country went into the war united. If this had not been the case it is unlikely that the enormous sacrifice demanded by the war would have been politically feasible.

This unity of purpose was only a part of Grey's achievement, though the one with the most immediate consequences. Of more importance for the future was the image that he left behind. It was essential that the British people should not think that they had been led into war by a rogue or a charlatan, a crank or an incompetent. On the whole they did not, and *The Times* probably reflected a widely-held view of Grey when it referred to him in the ponderous language of its obituary column as '... the diplomatist thanks to whom the pure and honourable character of our motives for the most momentous decision of a hundred years was made generally manifest'. *The Charm of Birds* was as much a part of this image as anything else that Grey did or said.

CHAPTER IV

GREY'S PLACES

FALLODON AND NORTHUMBERLAND

Fallodon is a small country estate in the far north of England, about twenty-five miles from the nearest part of Scotland. When Grey inherited it from his grandfather in 1882 it consisted of an eighteenth-century hall set in extensive wooded gardens, a Home Farm of about 200 acres, and various outlying farms let to tenants. All in all, Grey was the master of some 2,000 acres of Northumberland. He was thus well-equipped to lead the life of a country gentleman and that, when politics allowed him to get away from London, was largely what he did at Fallodon. As a young man he rode to hounds, fished the streams for trout and reared partridge and pheasant for the shoot, although these activities were never an important part of his life at Fallodon, and became less so as he grew older.

Grey is more closely associated with Fallodon than with

any other place. He regarded it as his home all his life. It was where he would like to have been born, as he once wrote. It was there that he died and that his ashes were buried with those of Dorothy.

Before it passed to Grey, Falloden had belonged to the Greys for over a century and a quarter. having come to them by marriage in 1755. The second Earl Grey, the famous 'Lord Grey of the Reform Bill', had been born there in 1764. However, when a splendid new mansion was completed at Howick in 1802, Falloden had been relegated to house a junior branch of the family, to one of which Grey belonged.

Of course, Fallodon was much more than an ancestral seat. Grey was very much involved in local and regional affairs and despite the impression which he often gave in London of being reserved, if not reclusive, he was remembered in the North East as widely known and greatly liked. This is not surprising — he represented Northumberland in Parliament for over thirty years, and was a director of the North Eastern Railway which played a key role in the regional economy. To some extent Fallodon was what would nowadays be called a power-base as well as a home.

The inspiration for *The Charm of Birds* was drawn from the Cottage, the New Forest and Wilsford in addition to Northumberland, but it is hard to imagine a more appropriate setting for its author than Fallodon. Positioned in a strip of rich arable farmland between the rolling heather of the outlying Cheviots and the blue expanses of the North Sea, it offers a variety of bird habitats within a short radius which it would be difficult to equal anywhere else. The upland moors support a distinctive population of small game birds (black grouse and partridge) and large birds of prey (the peregrine and hen harrier are known there). The moors are drained by fast-running streams (home of wagtail and dipper) which at their seaward end provide mudflats for the redshank, curlew and dunlin. The countryside around Fallodon itself is typical English farmland, gently undulating and divided into large fields by thick hedgerows which are themselves an important bird habitat for wren, dunnock, chaffinch and others. There is a generous amount of tree cover, much of it deliberately planted to shelter game or to enhance some farmhouse or hall.

A distinctive feature of Fallodon is its closeness to the main London to Edinburgh railway line which passes to the east. This was important to Grey, a frequent traveller to and from the English capital. Under the agreement by which the railway was built Grey had the right to stop trains at a small private halt which was virtually at the bottom of his drive. As a director of the railway company he was naturally interested in its management. He took an almost boyish pleasure in the great steam-hauled trains thundering past — it was a favourite joke to lure an unsuspecting guest to the side of the local signal box and watch him jump when the Scottish express, whose approach was hidden by the building, rushed past.

Fallodon Hall is approached from the direction of the railway by a long, tree-lined drive and another even longer drive runs off to the west down a mile-long avenue: the trees are indeed one of the most impressive features of the gardens. Grey himself was particularly proud of a large silver fir; a photograph of him standing in front of it appeared in *Twenty-Five Years,* his book about his work at the Foreign Office. There are also sequoia, deodar cedar, cedar of Lebanon and a large tulip tree, as well as many fine oak and beech.

The present Hall owes much to Grey, for he rebuilt it after a disastrous fire in 1917. He reused the brick and stone of the original and followed the original style, but he reduced the three storeys to two, thus greatly improving the proportions of the building and making it easier to run. The entrance to the Hall is dignified rather than imposing — with true Grey reserve its grander aspects are kept for the lawns on the south and west sides rather than displayed ostentatiously to the casual visitor coming up the drive.

A large skylight in the centre of the roof lights the central hall, along two walls of which runs the main staircase. Off this hall open the main rooms. They are well-proportioned and comfortable, in marked contrast to many other country houses. In the north-west corner of the ground floor is Grey's study, or the 'squirrel room' as it was known. Here the wild, red squirrels from the garden would come and keep Grey company as he worked, entering and leaving through an open window.

To the east of the Hall is a very large walled garden. This achieved some botanical renown long before Grey's ancestors came to Fallodon, but in his day its main purpose was to produce a useful quantity of fruit and vegetables. To the west of the Hall a long, flat lawn (perhaps the traditional venue for tennis and croquet) leads off to the site of Grey's aviary.

However Fallodon's most important features for Grey, both personally and ornithologically, lie on its southern side. Immediately outside his library he made a flower garden in memory of his first wife, Dorothy, and beyond this lies a second in memory of his second wife, Pamela. Both these places meant a great deal to him.

On the far side of the two flower gardens the ground slopes down to a small lake (Grey himself always referred to it as a pond). Constructed as an ornamental feature in the 18th century, it was adapted by Grey in the summer after he was sent down from Oxford for the breeding of waterfowl. This was a favourite hobby for the rest of his life. The remains of his 'fox-proof fence' are still discernible in the undergrowth. The bench where he used to sit, listening to his ducks if he could not watch them, still exists. In 1886 Grey built a second, smaller pond for his waterfowl below the original one, but this was not a success and the water has long since seeped away, although the site is still visible.

Until he retired from politics in 1916 the breeding of ducks at Fallodon was a pastime which Grey had to pursue vicariously, through his gardener, Mr Henderson. In *The Fallodon Papers* Grey said that to Henderson's '... interest in the birds, and the great care he has taken of them, is due the credit of such success as has been attained in the rearing of the different species'. Over the years no less than twenty-six species of duck had been successfully bred at Fallodon, ten of which were native species such as mallard, teal and pintail and thirteen were foreign, such as Chiloe wigeon and mandarin. They are listed in full in the passage quoted from *The Fallodon Papers* in Chapter XI. The 'duck dinner' when the ducks were fed at dusk each day either by Henderson or Grey himself must have been quite a performance: Grey would sometimes have to postpone or interrupt his own dinner for the occasion. He found that the unfamiliar sight of his evening dress — the white tie and tails which were still part of the English gentleman's evening routine — disturbed the ducks despite his encouraging calls and gestures; it was best to wear a coat. Sadly, without the regular feeding and protection from predators the lake at Fallodon no longer holds any particular attraction for ducks. The foreign species have long since disappeared, while of native species, usually no more than a single pair of mallards breed there.

Besides its connection with Grey, Fallodon was the birthplace of the Prime Minister who laid the foundations of parliamentary democracy (the second Earl Grey) and the home of the Home Secretary who is credited with having saved Britain from revolution in 1849 (Grey's grandfather). Yet its importance as an historic site has never received any official recognition. It is fortunate that Grey's successors have always done their best to preserve its links with the past.

Grey had no heir and he left Fallodon to a nephew, Mr (later Sir Cecil) Graves. The latter subsequently sold it to the late Colonel the Hon. H. Bridgeman, a Northumbrian who had known Grey. It now belongs to his son Mr Peter Bridgeman. The memory of Grey is warmly cherished and survives more vigorously in a house which is lived in than it might do in the dry atmosphere of a building which has become 'just' a museum. Many of Grey's small touches remain, despite the inevitable and necessary changes. His tall, tweed-suited figure may have passed forever, but at Fallodon it never seems to be far away. One half expects to see him striding off down a path or along a corridoor. Indeed, his personality is so indelibly imprinted on the place, in many small details, as to be almost as much a physical fact as the bricks and mortar, the trees and grass.

THE COTTAGE

During the best and happiest years of Grey's life the most important place in his world was the little wooden shack which was referred to somewhat whimsically as 'the Cottage'. This came about through his passion for dry fly fishing. As a schoolboy he had learnt the art on the stretch of the River Itchen which flows through Winchester, but it was to be practised to better effect further upstream where, as it happened, it crossed the land of his cousin, Lord Northbrook. When Grey came to London as a young MP he began to spend weekends fishing this stretch of the river. To begin with he and Dorothy stayed in the little village inn at Itchen Abbas. Later on, they had borrowed a fishing lodge belonging to Lord Northbrook. Finally, in 1890 Grey had the Cottage constructed just a few yards from the banks of the river.

It was a single-storey wooden building on brick foundations with a brick chimney and corrugated iron roof. It was soon covered in a great mass of ivy, honeysuckle, Virginia creeper, rose and wild clematis, so as to become very much part of the natural surroundings. Late Victorian second homes were not usually either so modest or so restrained. W. H. Hudson noted with approval that there were '... no gravel walks; no startling red geraniums, no lobelias, no cinerarias, no calceolarias, nor other gardener's abominations to hurt one's eyes and make one's head ache. And no dog, nor cat, nor chick, nor child — only the wild birds to keep one company.'

The Cottage was built at the end of an avenue which led to nowhere in particular from the main road to Winchester. On the other side of the avenue was a small chalk quarry which had become a favoured nesting site for many species, Grey had this fenced off as a miniature bird sanctuary. Perhaps the most distinctive feature of the Cottage was its location, which Grey must have chosen with some care. Although it is only a few hundred yards from Itchen Abbas and Lord Northbrook's large country residence, Avington House, it is completely secluded.

Grey described a typical weekend trip down to the Cottage in *Fly Fishing*:

> If our work will let us escape on Friday evening, it is luxury; but even if we belong only to those in the middle state of happiness, who work till midnight or later on Friday, and can have the whole of Saturday and Sunday in the country, we may still be splendidly well off, provided that we are careful to miss nothing. The earliest trains leave Waterloo, the usual place of departure for the Itchen or Test, either at or just before six o'clock in the morning. To leave London it is possible once a week, even after late hours, to get up in time for these early trains, and if you have no luggage (and you need have none if you go to the same place week after week), you will not find it difficult to get to the station. There are places where hansoms can be found even at these hours of the morning; they are not numerous, and they seem quite different from the hansoms that are abroad at more lively hours, but they can be found if you will look for them at certain places. The best plan, however, is to live within a walk of Waterloo, and as you cross the river in the early summer morning, you may feel more reconciled to London than at any other time, and understand Wordworth's tribute to the sight from Westminster Bridge. I pass over the scene at

Waterloo station, which at this hour is very different from the usual one, and the journey, on which perhaps one sleeps a little, though I have found that, while it is very easy to sleep sitting up in the late hours of the evening, it is necessary to lie down, if one wishes to sleep in the early hours of the morning. At some time between eight and nine o'clock, you step out of the train, and are in a few minutes among all the long-desired things. Every sense is alert and every scent and everything seen or heard is noted with delight. You are grateful for the grass on which you walk, even for the soft country dust about your feet.

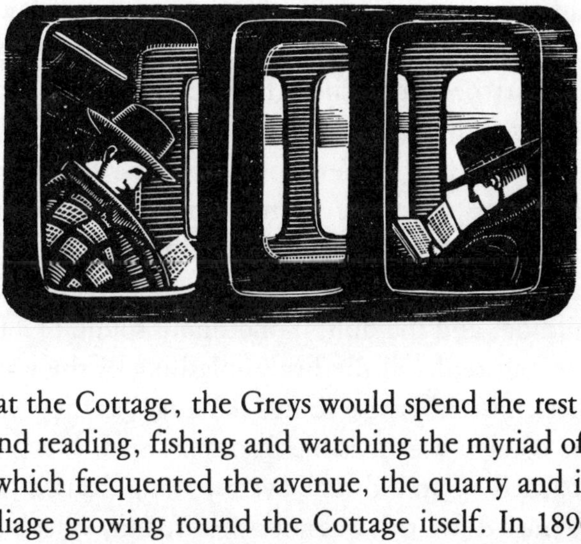

Once at the Cottage, the Greys would spend the rest of the weekend reading, fishing and watching the myriad of small birds which frequented the avenue, the quarry and indeed the foliage growing round the Cottage itself. In 1896 they brought a pair of bicycles down, so that they could make the occasional foray into the surrounding countryside. One such trip is described in an extract in Chapter IX..

In this way the Cottage came to have a special part in Grey's life, particularly in his relationship with Dorothy. In a retrospect which he wrote for the 1930 edition of *Fly Fishing*

Grey looked back on these earlier, happier days and considered the changes which had taken place around the Cottage since then.

> The cottage that I put up by the Itchen in 1890 was intended only as a fishing cottage; a place in which to get food, sleep, and shelter when I was not fishing. It became a sanctuary. The peace and beauty of the spot made it a sacred place. The cottage belongs to angling memories, but the fishing became a small part of the happiness that was associated with it. For thirty-three years the chosen spot remained a place of refuge and delight, not in the fishing season only. For the last four years, indeed, I had been unable to fish with a dry fly, and the original purpose for which the cottage had been put there had ceased to be. Great changes, however, had been taking place that were inseparable from a new epoch. For the first fifteen years there was little change and had been little change for many years before this time. I had seen the old mill at the village not far away replaced by a new building, and the dull, monotonous sound of a turbine had replaced the lively splashing of the water-wheel; but otherwise things remained as they were. The cottage was invisible from any road; it was approached by an old lime avenue, long disused, and the track down this was not suited for any wheels but those of a farm cart. There was a little wayside station on a single railway line close by; but the quickest route from London was to go by a fast train to Winchester, and thence to drive a distance between four and five miles to the nearest point to the cottage that was accessible by wheels. This was a drive of at least

half an hour in a one-horse fly. Presently taxi-cabs took the place of the horse conveyance and reduced the time of the drive to a quarter of an hour. Was this an advantage? On balance it was not. For escape from London meant that hurry, noise and bustle had been left behind: I had entered into leisure, where saving of time was no object, and often I would walk from Winchester to enjoy the country. There was a footpath way on each side of the river. By one of these one entered the cottage without, except for the momentary crossing of one road and of three secluded lanes, having had touch or sight of a road. There were thirty-three stiles on this path. It happened not infrequently that I could not get to Winchester till the latest train arriving there some time after eleven o'clock. The walk then lasted well into the midnight hour. In the dusk or dark it was easier to walk by the road than by the path. There was much charm in this midnight walk. Traffic had ceased, cottage lights had been put out, the inmates were all at rest or asleep. Now and then one heard in passing the song of a nightingale or a sedgewarbler, but in the main there was silence. It was pleasant after the hardness of London streets and pavements to feel the soft dust about my feet. On a still summer night there were sweet and delicate scents in the air, breathed forth from leaves and herbs and grass, and from the earth itself. It was as if one's own very being was soothed and in some way refined by the stillness, the gentleness and the sweetness of it all.

Then came the age of motors and tarred roads. Few people, I imagine, seek the smell of tar for its own sake. To me there is nothing unclean or nauseous in it, but it is a coarse, rough smell. The sweet and delicate scents of the night were obliterated by it, as if, overpowered and repelled, they had sunk back into the leaves and earth from which they had ventured into air. The strong smell of the tar seemed to disturb even the stillness of the night: the soft dust was no more, and the road was hard as a paved street. Not all, but much of the charm of the night walk was gone. There were other changes too; small houses of the villa type were built along the road that was nearest to the cottage: doubtless there are more of them now, for the cottage was accidentally destroyed by fire in January 1923, and I have not seen the place for some years. The sense of change was in the air. It may be that change is for the good:

> The old order changeth, yielding place to new,
> And God fulfils himself in many ways,
> Lest one good custom should corrupt the world.

It is not for us, who cannot foresee the future, who perhaps cannot rightly understand the present, to chide or to repine too much. Only it is impossible for us, who in our youth gave our affections to things that are passed or passing away, to transfer our affections to new things in which a new generation finds delight.

The beauty, however, of chalk-stream valleys still remains wonderful. The river still waters meadows that are unspoilt and unchanged, and its clear purity is guarded and protected.

> Still glides the stream and shall for ever glide,
> The form remains, the function never dies.

The remarkable thing is that since Grey wrote these words not a great deal has in fact changed around the Cottage. It is true that the new 'villas' have intruded a little on the seclusion of the place, but not as much as Grey might have feared. Maybe this is because others have found their sanctuary, their sacred place where he once did. The brick chimney still stands amidst a mass of greenery, the old avenue is still there, so too are the birds.

LONDON

Grey was born at 61 Chester Square in what was then considered to be Pimlico and is now thought of as Belgravia. Many years later he wrote 'I have always felt that my six brothers and sisters had the advantage of me in that they were all not only brought up but born at Fallodon.' It was at Fallodon that Grey lived during the holidays from Winchester and Oxford. At the age of twenty-three politics brought him to London and kept him there for most of the next thirty-one years. As a newly-wed and very junior MP Grey lived in Hereford Square, in Kensington. His first government job as Under Secretary at the Foreign Office enabled him to move to Grosvenor Road, which runs along the bank of the Thames between the Chelsea and Vauxhall bridges. Subsequently he moved to Whitehall Court, between the Thames and Whitehall. Finally, when he became Foreign Secretary he lived at 3 Queen Anne's Gate, five minutes walk from the Foreign Office and ten minutes from the House of Commons. It was one of the sad ironies of Grey's life that although he greatly preferred the countryside he probably spent most of his time in the centre of the largest city in the world.

He expressed his lack of enthusiasm for life in London in one of the best passages in *Fly Fishing:*

> ... This is not the place in which to write of the deep human interests of London, of what great affairs have their centre and of what issues are discussed and decided there. All that follows is written without any thought of denying or minimizing the attraction of these things for men's minds; but there is an aspect of London which is inevitable and becomes most oppressive in hot June days. There is the aggressive stiffness of the buildings, the brutal hardness of the pavement, the smell of the streets festering in the sun, the glare of the light all day striking upon hard substances, and the stuffiness of the heat from which there is no relief at night — for no coolness comes with the evening air, and bedroom windows seem to open into ovens; add to these hardships what is worse than all, the sense of being deprived of the country at this time and shut off from it. Perhaps you own a distant garden, which you know by heart, and from which occasionally leaves and flowers are sent to you in London; you unpack these and spread them out and look at them, spelling out from them and recalling to memory what the garden is like at this time. There were the young beech leaves and the sprays of double flowering cherry in May, and now there come the first out-of-door roses and the first of other things, perhaps the flower of some special iris lately planted. You see these things, you know the very trees, bushes, and places from which they were taken; you know the very form and aspect which the beauty of the season is taking in your garden, and you have

the knowledge that it is passing away, that you are missing for all this year things which are dear to you, both for the delight of seeing them afresh each season and for many old associations of other years. At such moments there surges within you a spirit of resentment and indignation, kept in abeyance during the actual hours of hard work, but asserting itself at all other times, and you pass through the streets feeling like an unknown alien, who has no part in the bustle and life of London, and cannot in the place of his exile share what seem to others to be pleasures. Work alone, however interesting, cannot neutralize all this, because it is only partly by the mind that we live. Mental effort is enough for some of the satisfaction of life; but we live also by the affections and where out-of-door things make to these the irresistible appeal, which they do make to some natures, it is impossible to live in London without great sacrifice.

THE NEW FOREST

Grey's friend, W. H. Hudson, noted in 1903 that the New Forest was becoming increasingly popular with visitors:

> And year by year, as motoring becomes more common, and as cycling from being general grows, as it will, to be universal, the flow of visitors to the Forest will go on at an ever-increasing rate, and the hundreds of today will be thousands in five years' time. With these modern means of locomotion, there is no more attractive spot than this hundred and fifty square miles of level country which contains the most beautiful country scenery in England.

One of these new visitors, discovering the forest for the first time was Grey. Exactly when is not known, but most probably it was in 1905 either just before or just after he became Foreign Secretary.

The attraction of the Forest was obvious: it was somewhere he could go at weekends in the autumn and winter when there was no fishing on the Itchen and the journey to Fallodon took more time than he could spare from the Foreign Office. Brockenhurst, on the southern side of the Forest was on the main Bournemouth to London railway line and it was there that Grey used to stay. It is probable that Grey had chosen Brockenhurst as a weekend retreat in the event of his becoming Foreign Secretary at some stage before the eventuality arose. We know that it was Dorothy who, a short time before her death, found him a hotel there. Most probably this was the Forest Park Hotel which still exists today.

In *Twenty-Five Years* Grey described and justified a typical weekend in the New Forest.

> In autumn and winter I found a quiet hotel opening on to a heath in the New Forest. There I could have the same private rooms at the end of each week. On Sunday morning I might start between 11 o'clock and midday, walk off into the Forest, eat my pocket-luncheon in some wild part of it, drop into an inn at Beaulieu, Lyndhurst, or Burley for tea, and thence get back to the hotel on the outskirts of Brockenhurst in the evening, in the dusk in early spring, under moon or stars in winter. The other hours of the day were available for reading or work. Early on Monday morning I returned to London with all arrears of work done, lungs filled with fresh air, limbs stretched, mind and body refreshed. These week-end expeditions have sometimes, I am told, been questioned, as implying slackness in work; they were, it is true, planned for pleasure and not for duty, but they did in fact suit the work much better than any other way of spending the week-end out of London. The ordinary country-house visit so often means neither work, rest nor exercise; I made sure of all these, and the anticipation of these weekly escapes kept up my spirits during many weary hours of work in London.

WILSFORD

The little village of Wilsford lies in a valley scooped out of the downs of Salisbury Plain through which the River Avon flows on its way from Amesbury to Salisbury. Wilsford Manor belonged to Grey's second wife Pamela and he must have spent a considerable amount of time there in the years 1922-8.

In 1903 Wilsford Manor and the estate which went with it, had been purchased by Edward Tennant, who was Grey's friend and Pamela's first husband. The Tennant family had made their fortune in Scottish industry and then taken an interest in politics. From 1906 to 1910, when he was elevated to the peerage as Lord Glenconner, Edward Tennant was MP for Salisbury. It was perhaps this connection that determined his choice of Wilsford as a country seat.

The present manor house was built for the Tennants in 1904-6 on the site of an earlier building. It is an attractive

house of modern proportions built in the style of the 17th century with stone mullioned windows and walls of stone and flint chequer-work. About a hundred yards from the manor stands the little parish church in which Grey's second marriage was celebrated in 1922. Next to the church is a square Georgian vicarage. On the other side of the road are the half-dozen cottages which together with the church, vicarage and manor comprise the entire village. It is a quiet, almost secretive sort of place, which, one imagines Grey would have found very attractive. Indeed, the *Victoria County History,* relying probably on an oral tradition, states that until Pamela's death in 1928 Grey lived mainly at Wilsford. If so, a large part of *The Charm of Birds* must have been written there.

Grey would particularly have liked the extensive, richly wooded garden of the Manor. Pamela describes it as Grey would have known it in one of her poems: 'This garden has a soul, it has its moods,' she wrote, and one can easily picture Grey walking through '… the leafy silence of its woods that are so thickly grown with hemlock flower,' reliving memories of the past, trimming and polishing them in his mind for inclusion in *The Charm of Birds.*

PART II

CHAPTER V

ROOSEVELT AND THE 'BIRD WALK'

On the 14th of June 1910 Grey went on a long walk with a distinguished foreign visitor: Theodore Roosevelt, former President of the United States of America. Many years later he explained how this came about in a lecture to the students of Harvard:

> Several years ago when I was at the Foreign Office in London, I got a letter from Mr. Bryce, who was then British Ambassador at Washington, saying that President Roosevelt intended to travel as soon as he was out of office. He was going to travel in Africa, to visit Europe, and to come to England, and he was planning his holiday so minutely as to time his visit to England for the spring, when the birds would be in full song and he could hear them. For this purpose he

> wanted it to be arranged that somebody who knew the songs of the English birds should go for a walk with him in the country, and as the songs were heard tell him what the birds were. That is a pretty good instance of thorough planning in advance for a holiday. It seemed to me very attractive that the executive head of the most powerful country in the world should have this simple, healthy, touching desire to hear the songs of birds, and I wrote back at once to Mr. Bryce to say that when President Roosevelt came to England I should be delighted to do for him what he wanted. It is no more a necessary qualification for the Secretary for Foreign Affairs in London than it is for the President of the United States that he should know the songs of the birds, and it is an amusing coincidence that we should have been able to arrange this little matter satisfactorily between us as if it were part of our official duties, without feeling obliged to call in experts.

Roosevelt and Grey had much in common besides their love of birds. Both men had lost their first wives in tragic circumstrances. Alice Roosevelt's death in childbirth had been a great blow to Theodore. However, within two years he had been able to remarry and became very devoted to his four sons and two daughters. Again, both Roosevelt and Grey had a rather diffident view of the political success which they had achieved. Roosevelt, gratuitously and to the chagrin of his friends and colleagues, had renounced the possibility of a third term as president immediately after the start of his second term in 1904. Roosevelt, like Grey, belonged to the upper echelons of the society into which he was born: his family was one of the oldest and most pre-eminent in New York.

Of course in personality the two were quite different. Grey as we know was reserved, restrained and, as his critics would have it, altogether too gentlemanly. Roosevelt by contrast was ebullient, impulsive and very full of himself. His idea of a country estate was a cattle ranch in the Dakota badlands whence he would retire from the gentle civilities of New York to enjoy the rough perils of life as a cowboy. When America had gone to war against Spain in 1898 Roosevelt's response had been to abandon his post as Assistant Secretary of the Navy to become second in command of a volunteer cavalry regiment, the famous 'Rough Riders' (hence Grey always referred to him with the prefix 'Colonel'). The regiment's charge up San Juan Hill made him a national hero, and opened the way to the White House. As one of his close friends remarked when he got there, 'Never forget that the President is about six years old.'

Despite their different personalities Grey and Roosevelt got along famously. Besides their shared love of birds, both men were highly successful and highly adept at elective politics: they spoke the same language.

For Grey it was a good thing to be able to get on well with a man who still had considerable influence on American policy. But the main significance of the bird walk was as a symbol: two important national leaders walking quietly side by side towards a common destination in pursuit of a common interest. There are inevitable echoes of that 'special relationship' some thirty years later between Grey's cabinet colleague Churchill and Roosevelt's distant cousin Franklin Delano Roosevelt, which put paid to Hitler and in some measure thwarted Stalin. To Grey, in later years, memories of his walk with Roosevelt must have led to thoughts of the American support in the First World War which made Allied victory a certainty. This was partly due to his own careful handling of Anglo-American relations in the early years of the war.

Kaiser Wilhelm had happened to visit the New Forest in December 1907 one weekend when Grey was there. Grey apparently went to some trouble to avoid him. 'The German Emperor is perilously near, but I know he can't find me in the forest — people with motor cars are chained to the roads and there is no freedom like the freedom of one's own feet,' as he wrote to a friend.

Of course it is important not to read too much, by way of symbolism or otherwise, into the bird walk — it was only a weekend ramble undertaken by two middle-aged men. There is all the same a certain charm about it and, again, this is best described in Grey's own words, taken from his lecture to the Harvard students:

> Time passed, and when the President retired from office he went to Africa and had much big-game shooting and travel there. Then he came by way of the Sudan and Egypt to Europe. The leading countries of Europe were stirred to do him honour, England not less than others. He had a great reception and everywhere a programme of great and dignified character was arranged for him. European newspapers were full of it long before he got to England, and I thought this little walk to hear the songs of English birds suggested some two years before would be forgotten and crowded out by greater matters. But it was not so. Without any reminder on my part, I got an intimation from the English friend who was to be Colonel Roosevelt's host in London that Colonel Roosevelt had written to him to say that this promise had been made and

that he wished time to be found for the fulfilment of it. I saw Colonel Roosevelt once soon after he came to London. The day was arranged and at the appointed time we met at Waterloo Station. We had to ask the newspaper reporters not to go with us, not because it made any difference to Colonel Roosevelt, but because birds are not so tame, or perhaps I should say are more self-conscious than public men and do not like to be photographed or even interviewed at close quarters, and it was necessary, not only that Colonel Roosevelt and I should be alone, but that we should make ourselves as inconspicuous and unobtrusive as possible.

So we went alone, and for some twenty hours we were lost to the world. We went by train to a country station where a motor was awaiting us. Thence we drove to the little village of Titchborne in Hampshire, and got there soon after midday. In the village of Titchborne there lives also the family of Titchborne, and in the old village church there is a tomb with recumbent figures of one of the Titchbornes and his wife who lived in the time of James the First; on it is inscribed the statement that he chose to be buried with his wife in this chapel, which was built by his ancestor in the time of Henry the First. That shows a continuous record of one family in one place for some eight hundred years. I forget whether we had time to go into the church and look at it, but the songs of the birds which we had come to hear are far more ancient. They must be the same songs that were heard by the inhabitants of England before the Romans came, for the songs of birds come down unchanged through great antiquity, and we are listening today, in whatever part of the world we may be, to songs which must have been familiar to races of men of which history has no knowledge and no record.

I was a little apprehensive about this walk. I had no personal acquaintance with Colonel Roosevelt before he came to England in 1910, and I thought to myself, 'Perhaps, after all, he will not care so very much about birds, and possibly after an hour or so he will have had enough of them. If that be so and he does not care for birds, he will have nothing but my society, which he will not find sufficiently interesting for so long a time.' I had relied upon the birds to provide entertainment for him. If that failed, I doubted my own resources. I need have had no fear about his liking for birds. I found, not only that he had a remarkable and abiding interest in birds, but a wonderful knowledge of them. Though I know something about British birds, I should have been lost and confused among American birds, of which unhappily I know little or nothing. Colonel Roosevelt not only knew more about American birds than I did about British birds, but he knew about British birds also. What he had lacked was an opportunity of hearing their songs, and you cannot get a knowledge of the songs of birds in any other way than by listening to them.

We began our walk, and when a song was heard, I told him the name of the bird. I noticed that as soon as I mentioned the name, it was unnecessary to tell him more. He knew what the bird was like. It was not necessary for him to see it. He knew the kind of bird it was, its habits and appearance. He just wanted to complete his knowledge by hearing the song. He had, too, a very trained ear for bird songs, which cannot be acquired without having spent much time in listening to them. How he had found time in that busy life to acquire this knowledge so thoroughly it is almost impossible to imagine, but there the knowledge and training undoubtedly were. He had one of the most perfectly trained ears for bird songs that I have ever known, so that if three or four birds were singing together he would pick out their songs, distinguish each, and ask to be told each separate name; and when, farther on, we heard any bird for a second time, he would remember the song from the first telling and be able to name the bird himself.

He had not only a trained ear, but keen feeling and taste for bird songs. He was quick to express preferences, and at once picked out the song of the English blackbird as being the best of the bird songs we heard ... Colonel Roosevelt liked the song of the blackbird so much that he was almost indignant that he had not heard more of its reputation before. He said everybody talked about the song of the thrush; it had a great reputation, but the song of the blackbird, though less often mentioned, was much better than that of the thrush. He wanted to know the reason of this injustice and kept asking the question of himself and me. At last he suggested that the name of the bird must have injured its reputation. I suppose the real reason is that the thrush sings for a longer period of the year than the blackbird and is a more obtrusive singer, and that so few people have sufficient feeling about bird songs to care to discriminate.

One more instance I will give of his interest and his knowledge. We were passing under a fir-tree when we heard a small song in the tree above us. We stopped and I said that was the song of a golden-crested wren. He listened very attentively while the bird repeated its little song, as its habit is. Then he said, 'I think that is exactly the same song as that of a bird that we have in America'; and that was the only English song that he recognized as being the same as any bird song in America. Some time afterward I met a bird expert in the Natural History Museum in London and told him of this incident, and he confirmed what Colonel Roosevelt had said, that the song of this bird would be about the only song that the two countries had in common. I think that a very remarkable instance of minute and accurate knowledge on the part of Colonel Roosevelt. It was the business of the bird expert in London to know about birds. Colonel Roosevelt's knowledge was a mere incident acquired, not as part of the work of his life, but entirely outside it. I remember thinking at the time how strange it seemed that the golden-crested wren, which is the very smallest bird which we have in England, should be the only song bird which the great continent of North America has in common with us.

One trivial incident there was in our walk which gave us some amusement. We were going by footpaths down a river valley, a very beautiful, but a very tame and settled country, where anything like an adventure seemed impossible. We were on a path which I had known for many years, and along which I had walked many times, not only without adventure, but without even incident. Suddenly we found ourselves stopped — the path was flooded, some weeds had blocked the river close by, and instead of a dry path we had about twenty yards of water in front of us. The water was not very deep,

certainly not above our knees, but I had not intended that there should be any wading in our walk nor had I prepared for it. I asked if he would mind going through the water, to which, of course, he replied that he would not. So we went through, got wet, and in the course of the afternoon got dry again as we walked. Nothing of the same kind had happened there before; nothing has happened since. I think there was some magnetism about Colonel Roosevelt's personality which created incidents.

After going a few miles down the valley, we got into our motor, which was waiting at a village inn, and drove to what is called the New Forest, though it is more than eight hundred years old. We were now in a country of wild heath, quite uncultivated, and the part we went through was mostly natural forest. Here we heard some birds different from any we had heard in the valley of the Itchen, and got to a little inn standing on the open heath about nine o'clock in the evening. We had dinner, and next morning we breakfasted together and went to Southampton, whence Colonel Roosevelt returned to America.

I am not attempting here a full appreciation of Colonel Roosevelt. He will be known for all time as one of the great men of America. I am only giving you this personal recollection as a little contribution to his memory, as one that I can make from personal knowledge and which is now known only to myself. His conversation about birds was made interesting by quotations from poets. He talked also about politics, and in the whole of his conversation about them there was nothing but the motive of public spirit and patriotism. I saw enough of him to know that to be with him was to be stimulated in the best sense of the word for the work of life. Perhaps it is not yet realized how great he was in the matter of knowledge as well as in action ... Not long ago when an English friend of mine was dying, his business agent came over to see him. One of the family asked the agent whether he had come on important business. 'No,' he said, 'I have come for a little conversation because I was feeling depressed this morning and I wanted to be made to feel two inches taller.' That saying would, I think, have been specially applicable to Colonel Roosevelt also. He could make people feel bigger and stronger and better.

CHAPTER VI

EARLY SONG

THE ROBIN

First let the robin be noted. He has been singing since August, whenever the weather was not unduly discouraging, and he will go on singing till July. W. H. Hudson told me that the female as well as the male robin sings: this seemed probable, for in autumn each robin is alone in its territory. No robin will then tolerate a companion of its own or even of the other sex. Yet in each territory there seems to be song. The observations of Mr Burkitt with ringed birds, of which the sex had been ascertained and could be identified, has proved that female robins do sometimes sing like the males. I have, however, never heard the female sing after the birds have paired for the breeding season: of the pairs that I have had under observation only one bird has sung, and I conclude that this has been the male. Certainly only one bird has sung at a time and one has been silent, when they were together. If any one suggests that it is sometimes the male and sometimes the female that sings, I can only say that I think this is unlikely, but I admit that robins are capable of anything.

Is there a difference between the quality of robin song in autumn and in spring? I think there is. In autumn the song has something thin and acid in its tone. 'The bitter note of the robin,' was the comment of a friend, as we passed close to a bird singing in October. In spring the song seems more vigorous: it is worth while to stand close to a good robin and listen attentively: some notes of fine quality will be heard. In April, when thoughts are turned towards summer warblers, I have even heard one or two notes in a robin's song that prompt the exclamation 'blackcap!' In estimating the difference between spring and autumn songs allowance must be made for the human mood and expectation of the mind. In autumn, when

> The warm sun is failing,
> The bleak wind is wailing,

when

> The chill rain is falling,
> The nipped worm is crawling,

and the sun is getting lower and the days shorter, our own minds are attuned to a minor key, and we find it in the robin's song. On a warm April day, when sap is rising and we are full of anticipation, with ears a-tiptoe for the first note of a blackcap, we judge the robin's song differently. 'We used,' said a Conservative who was cutting my hair soon after the war, 'we used to think Mr Lloyd George everything that was bad. Now we admire him. Is it he or is it we that have changed?' And so I ask, listening to a robin in spring and comparing the impression remembered of the autumn, 'Is it the song or is it I that have changed?'

Be this as it may, the robin's song is worth attention: he sings more than any of our birds; he may be heard in every month of the year, even in July and August, if we listen for him: and, though he may not open the Great Chorus at Dawn in May, he is the last to cease in the evening, outstaying even the thrush.

THE BLACKBIRD

Let us imagine a perfect day near the end of February: a quiet day, the sun shining, the air mild; we get a few such days in every normal February. A lover of birds is walking in the bare woods, pleased with the various songs that he has already heard this year, listening especially to the loud song of dominant mistle-thrushes, fresh as the bright air, clear as the blue sky above; while little rills of robin song pour into his ears. Suddenly attention is arrested by something more rare in quality than the song of the mistle-thrush; he stops, listens, hears it again, and is assured of the authentic notes of a blackbird. I have told elsewhere how the trained and sensitive ear of Theodore Roosevelt selected the song of the blackbird as the best of all the songs that he heard on a certain day in early June in England. It is not possible to explain why the blackbird's notes excel and why they mean so much to us. To me there is something in it that I can best describe as intimacy. The songs of other birds please or delight us, but that of the blackbird seems to make a direct appeal to us and stirs some inward emotion. It is one of five bird sounds for which I have a particular affection. The song is a linked phrase, repeated again and again at intervals. As a rule the bird chooses a perch for the purpose of singing, but occasionally changes its place and sings while flying from one perch to another. It is a supreme moment when a blackbird, passing from one bush of flowering hawthorn to another, as if one spot could not contain its joy, utters its song on the wing in the ecstasy of mid-May.

There is a general type of song common to all blackbirds, but sometimes one individual bird develops a phrase peculiarly its own, and can be identified by this in the same place day after day throughout the season of song.

Blackbirds have a common trick of ending their song with an inferior note, almost a squeak. John Morley in his *Recollections* describes a visit he made with a friend to Renan. We are told how the old man was roused to interest, and we are made to feel the vigour and magnificence of his talk. Morley and his friend had planned to go to the opera, but with the glow upon them of what they had seen and heard of Renan, they felt that the opera would be bathos, and they did not go to it. Then Morley adds as an additional reason for not going to the opera, that, besides, he had to pack. The squeak or 'scritch' at the end of a blackbird's song, coming after the beautiful and fine notes that precede it, is like that — a sudden drop from heights.

A short four months of song is all that a blackbird gives us in the year. Indeed, he is not established in song till March, and he ceases before the end of June. Then he goes into his moult and becomes an untidy and deplorable object.

THE STARLING

The starling's song (for his very musical performance must be given that name) has been heard all through the autumn. When thrushes do not sing it is the most valuable and conspicuous event in bird music during the autumn and winter: every evening about sunset a starling, or a little party of them, sits on the top of a bare tree and gives a variety of performance; some notes are their own, others are plagiarisms more or less close to the original; some are perfect imitations. The note that seems to me to be peculiarly the starling's own is a very clear boyish whistle; with this are all manner of other sounds — poultry noises, the chatter of sparrows, the cry of a peewit — one never knows what is coming next. Some of the notes are very pleasant, even beautiful, and are a delight to hear. The starling cannot attain to the quality of a blackbird, but he can whistle so near to it that more than once in February have I stopped and listened, thinking that I had heard all unawares the first blackbird, till some chatter or chuckling satisfied me that it was only part of the starling's infinite variety. Where curlews breed, the spring notes of the curlew are a very favourite imitation. Nobody but a curlew can make that wonderful sound; but this does not deter the starling from attempting it, and the resemblance of the imitation to the original is sufficient to make us thank him. The starling is like a gramophone among bird songs, and it has chosen some of the best of which to make records.

Last September (1925) one bird was missing from a tame covey of young partridges that I was feeding. From another part of the garden came the call of a partridge so perfect that I went hopefully in search. On one or two other evenings I heard the same call frequently repeated, but there was no living partridge to be found, and it was no doubt the trick of a starling (there were some singing at the time), though I did not catch the bird in the act of deceiving me.

Close to the house at Fallodon, standing single on the lawn, is an old elm tree. It is too large for that part of the garden: it is failing and becoming more unsightly every year, for some of the branches at the top are dead. Yet the tree is spared because those same dead branches that seem to dishonour the tree do in fact gain distinction for it. In September and October, while the tree is still thick with leaves, these bare dead branches at the top are the favourite assembling place of a small party of starlings. There they sit every evening about sunset, and one or more birds discourse. It is worthwhile to sit and listen to them, not for the interesting variety alone, but also for the beauty of some of the sounds they make.

THE THRUSH

The thrush has a variety of notes, but the order in which he gives them is improvised. We may listen to a thrush for a time without hearing the notes we most desire, for some of his notes are much less agreeable than others; a musical phrase resembling 'did-he-do-it?' may be repeated two or three times and then abandoned for some other notes. In fact, the manner of the thrush when singing gives an impression of selection and choice of the sound that he will make next. He sings perched in a tree, to which he has mounted for this purpose. There he will maintain his position and his song for some time, especially about dawn and sunset, preferably on the same tree day after day, pausing in this performance as if to select and choose his notes. Probably if birds were to be regarded as endeavouring to please us by song, the thrush should be put first among British birds. He does not rank in the very highest class for quality, but he certainly comes high in the second class. He is undoubtedly a major song, and owing to the number of thrushes, their persistent singing and the many months in which they are to be heard, we hear more of their song in the South of England than that of any other bird, except the robin. In song the thrush seems to be working very hard to please, and he succeeds. His song, too, can give a very pleasant impression of quiet contentment as well as of exultation.

Occasionally a thrush will introduce some freak sound and make it part of his song. Many years ago an attempt was made to keep white-faced whistling ducks in the collection of waterfowl at Falloden. They are not hardy, but one of them survived for about two years, and being very tame, as is the manner of its kind, the whistling note with which it saluted every one it saw became the most distinct and familiar sound in the garden. In January it died. In the following April we came home for Easter, and from high up in a silver fir by the pond came a perfect imitation of the call of the white-faced whistling duck. It was made by a thrush, but the bird did not continue the imitation after this spring, though it was often to be heard during that Easter holiday.

Another example was that of a thrush in the garden at Wilsford in the Avon Valley. For four seasons, at least, it sang continually during May and June on two notes; they were none of the usual notes of thrush song: they were as monotonous, but louder or more staccato than those of a chiff-chaff; and the persistence and monotony of them were wearisome. Nevertheless we listened for them every season in the same region of the garden. Last year (1925) my wife and I ascertained by close observation of the bird that he could and sometimes did sing the ordinary song of the thrush, so that his preference for loud monotony was due, not to incapacity, but to perversity. His freak song was not an imitation of any other that I know. So remarkable and well known did the sound become that the bird came to be spoken of in the family as 'Monotone'. We never heard him in the autumn or winter, but in spring the first hearing of 'Monotone' was noted, welcomed, and announced with as much satisfaction as the first news of a chiff-chaff or willow-warbler.

THE GREAT TIT

The great tit must have his place among the songbirds of this month [January]. His spring notes begin to be heard now. The rhythm of the notes suggests sawing, but they are evidently intended by the bird to be a song as much as that of any other bird that sings. I have known them to be mistaken early in the year for those of the chiff-chaff, before the latter bird has arrived.

The sound contribution of the great tit to the bird sounds of this month is no mean one. The notes ring out loud, vigorous, and clear: to one who knows the ways of the great tit the sharpness of his notes suggets the sharpness of his beak. Among others of his own size he is a strong, bold bird,

capable of tragedy, as the following story will show. There were traps kept in the garden for rats and other small nuisances. Some of these traps were cages so constructed that it was easy from outside to find the way in, but difficult from the inside to find the exit. Into one of these there had entered a dunnock and a great tit. Presumably the dunnock had entered first, and the tit had come later, attracted by seeing the cage occupied. Whether the tit entered with fell design, or whether having entered he was roused by what Shakespeare calls 'vile opportunity', we cannot tell but the result is known. When in due course I visited the trap the dunnock was a pitiful sight: it lay dead; the skull was broken into and the brain had been eaten. The great tit alone was alive, a patent and thriving murderer.

'What did you do with the horrible tit?'

'Madam I set him free, not feeling competent to assess his moral responsibility in the matter.'

After all, do we not most of us follow the advice or instruction to 'kill and eat', or eat what has been killed for us?

THE MISTLE-THRUSH

The mistle-thrush may be heard in mild winters before January, but I think of his song as generally beginning in that month: to be listened for then as a sign of reviving life in a new year. He is a bird that occupies a fixed station; that takes up a 'stance' for the express purpose of singing. This is one distinction that may be noted in the habits of song-birds: some sing from a fixed place, others sing as they go about their business of the day. Of the birds [already] mentioned the robin belongs to the first category, the wren, I think, mainly to the second, as do also the summer warblers.

The mistle-thrush mounts to the highest convenient stance he can find: he will resort to this same place day after day, like the song-thrush.

His song suggests to me a linked phrase, and here we come upon another possible classification of song-birds — those that, like the mistle-thrush and blackbird, express themselves in a linked phrase; repeated, after a pause, with slight, if any, variation: those whose song suggests rather a sentence than a phrase; this is repeated at intervals and is, as a rule, the same time after time. Of this the willow-warbler is a typical example. Thirdly, the birds whose song seems to be an improvisation of separate notes, not unlimited indeed in variety, but so indefinite in arrangement that we cannot be sure what notes are coming next. The song-thrush is the most complete and familiar example of this type of song.

The mistle-thrush is best appreciated in January and February. Not that he sings better then than later on; indeed his song is more perfect in April, but it is very good in the two first months of the year, and it stands out clear in the then comparatively silent air. There are boldness and wildness as well as sweetness in the tone. It has not the rich and moving quality of the blackbird, and yet it stirs us. For on a windy day in January, when the blackbirds seek the shelter of laurels and thickets and have not a note of song in them, the mistle-thrush sings, aloft and conspicuous. There is, it has been well said, 'weather in his song'. Birds as a rule seem to dislike wind more than any other sort of weather, but the mistle-thrush is less discomfited by it than any other song-bird. On a windy day early in the year the 'storm-cock' will mount his tree, and there in full exposure proclaim by song that he is vigorous and glad. Every year he deserves to be honourably and gratefully saluted.

THE DIPPER

The dipper ... is the most certain January singer, for even the hardest weather does not silence him. When the woods are hushed and white with snow, and the burn is pinched by frost, so that only a narrow dark channel of running water shows between the ice and snow at the side of it, there on some stone in the burn the dipper will stand and sing. It is water rippling over a stony bed that he frequents; the soft luxuriance of a chalk stream has no attraction for him. His song seems part of the sound of the rippling water, from which he is never away. 'I hear thee where the waters run' may well be said of the dipper. His song is very sweet and lively; it has no marked beginning or close, but goes on indefinitely. It is as if 'beauty born of murmuring sound' had passed into the bird who was giving it back as song to the stream whence it had come. I reckon that there are two pairs of dippers each with its own length of the little burn at Fallodon; and where a burn is narrow and the banks upstanding, one can approach close to the sound and listen with pleasure and with admiration of the birds' hardihood. Of this hardihood I had experience in the first week of a certain memorable March in Sunderland. Blizzard followed blizzard; feet could make but slow progress through the snow; wheels could not travel except where passage had been cut by man through the drifts, on a small section of road, and even this was blocked again by fresh blizzards, that obliterated man's puny efforts. The frost was intense; the river was frozen from bank to bank, except where the swift current kept some open water at the head of pools. Wild life was helpless: sheep had to be searched for and dug out of snowdrifts: grouse in trouble and despair flew in bewildered packs about the white hillsides. At a little height above the river my friend and I were slowly making our way on foot through the deep snow. From an unfrozen stream below there came up to us the sound of a dipper, singing its full song, undeterred by the conditions that were distressing all other life, unaffected by the cold, undismayed by the desolation. It was another moment when the song of a single bird penetrates to the affections and abides thereafter in the memory.

THE WREN

The wren's song is a succession of rapid notes, forming a long musical sentence, that is repeated again and again at intervals. The full sentence is a long one, but the bird very often begins it and leaves off in the middle or even after the first few notes: a good example of what we were taught at school to call 'aposiopesis'; like the woman who, after speaking for a time at a public meeting, began a sentence with the words 'But still ...,' then stopped and sat down. But when a wren is in good form he sings, as it was said the young Queen Victoria danced at a Court function in Paris, 'with decision, and right through to the end'.

To appreciate the song, however, the person of the wren, as well as its voice, must be taken into acount. The song is a good loud one, but when considered in relation to the tiny body, it is positively mighty. 'Valiant', 'resounding', are epithets that I have applied when listening to a wren singing: and a little boy, whose mother had taught him to be familiar with the song, spoke, while doing lessons in an open-air parlour, of 'that shattering wren'. Great as the effort of the song seems to be, a wren at the height of the season will repeat it at short intervals for a long time without tiring.

There are individual wren songs that stand up like little peaks in memory.

Once, in days of youthful ignorance, a loud song proceeding from a thick cypress arrested my attention. It was long ago, in March 1884. I wondered what bird this was, and struck the branches with my stick. Out flew an unabashed wren, to perch elsewhere, and repeat the singing. The contrast between the size of the body and the strength of the song impressed itself on me, then and there, never to be forgotten.

There was another occasion, some years later. We had a cottage in the Itchen Valley, and had gone there early on a Saturday morning after a weary week in London; it was about 8 o'clock on a fine warm morning. I had just arrived, and stood in the doorway that opened on to the little lawn. Escaped from London at this season of the year, on such a day and in such a place, with the weekend prospect before me, I was indeed 'standing on the top of golden hours': in front, some ten yards aways, was a poplar tree, and from it a wren sprang into the air, and, singing in an ecstasy as he flew, passed straight over me and over the cottage roof to some other place of bliss on the farther side: 'like a blessing', said one who was with me.

Wordsworth in the 'Prelude' records an incident, differing in mood and setting, but similar in kind, in wich the song of a single wren animated a ruined chapel. This was remembered, and added to that store of 'emotion recollected in tranquillity', which he gives as a definition of poetry.

A third memory is of recent date. The incident occurred in the greenhouse, where a wren was in the habit of coming and going through open windows or ventilators. I heard the wren singing with even more than usual violence, perched somewhere among the plants. The cause was plain — another wren was singing outside, and there was a song combat. When two birds in neighbouring territories are singing in rivalry, it is the rule that they do not sing together: each in turn listens while the other sings; then the bird that has listened replies. So it was now. The greenhouse wren was used to seeing men about the plants, but besides this familiarity, it was so intensely occupied in listening to the song outside and then exerting itself to the utmost in reply, that it took no notice of my approach. I watched it for some time at a distance of not more than two or three feet, perhaps even less. The strophe and antistrophe went on: the attitude of my wren when listening was intent and still; when it replied the animation and vehemence were such that it seemed as if this little atom of life might be shattered by its own energy.

Not always, however, does the wren's song give this impression of force. On a bleak day in autumn, when chill wind is blowing, a wren's song will unexpectedly be heard, and at a distance will sound thin. Wordsworth describes this aspect of wren song:

> To the wind she sometimes gives
> A slender, unexpected strain.

CHAPTER VII

INCREASING SONG

THE DUNNOCK

The dunnock this year (1926) I did not hear till February; then its song was soon established. The dunnock is quiet in colour and in manner, and in its demeanour towards other birds it gives an impression of meekness. A friend to whom this remark was made replied, 'Yes: but then it relieves its feelings by laying that wonderful egg.' It relieves its feelings also in song. The song is quite adequate in loudness, and is pleasant, but there is little shape or feature by which to remember it. I found it a difficult song at first to distinguish, till I became familiar with the 'timbre'. It is, in fact, by the 'timbre' that in the end we get to recognise bird songs and distinguish one from another with greatest certainty. There are many distinctive characteristics of songs, which have little or no resemblance to each other; but as years go on I think it is by the difference in tone that we tell most easily one from another when hearing several singing together: just as one may recognise the presence of a friend in company by the tone of the voice, without hearing or attending to what is said.

The dunnock's song has spirit and uplift, and he often chooses a conspicuous position on the top of a shed or of the garden wall. Otherwise he is unobtrusive. Seclusion in the stack of pea-sticks, collected for future use in the kitchen garden, is a favourite habit of the dunnock. It is not a shy bird; it frequents back premises of houses and the gardens behind them, and is easy to watch; its habits suggest not so much fear of man as an inclination to apologise for its presence.

THE TREE-CREEPER

The tree-creeper I have heard in January, but I hardly expect its song then. It is a very sweet little wisp of sound. At first I confused it with that of the gold-crest, and found it difficult to keep the two songs distinct. The distinction is now so clear that it is made automatically and without effort, and I can hardly say that one song now suggests the other to me. I have read in some book that the song of the tree-creeper is rarely heard. This is not altogether true. In many Aprils, when fishing a salmon river in Sutherland, with spruce, oak and beech woods on one bank, the songs of tree-creepers, as well as of gold-crests, have been heard, repeated frequently and abundantly for many days together. But it is true to say that the tree-creeper is a most capricious singer. Even on a most favourable day in April it is not safe to say, 'Let us go out, and we shall hear a tree-creeper.' On the other hand, he will surprise us by uttering his song in some month early or late, when we do not expect it. In 1915 the Russian Army was hav-

ing a terrible series of reverses: unbroken bad news came for a long period. One day came news of a distinct Russian success at one point in the line. Some one on hearing it remarked, 'You can never depend on the Russians not to do well.' So it may be said of the tree-creeper, 'You can never depend on him not to sing.'

He is very decidedly not a set singer, but desultory: that is to say, he does not sit in one place for the purpose of singing, but utters his song as he goes about the trunk of a tree searching the bark: the song is not a separate part of the day's business, but is incidental and subordinate to it. It is many years since my sight enabled me to observe birds by eye, except when very close to me, but my recollection is that the tree-creeper opens the beak very widely to sing. It is a very attractive bird; the slender bill suggests a humming-bird; the back is a rich arrangement of brown colours; the under parts are silvery white, and the body had a delicate outward curve from head to tail as it searches the tree trunks. It works up a tree and never appears head downwards. Thus, when it has done with one tree it flies to the low part of another and begins working up again. Has any one ever seen a tree-creeper still and at rest? It does not perch on twigs: it is always busy on the trunk or on the stem of some big bough. It does not seem to go into holes like woodpeckers. There are several Sequoia Sempervirens at Fallodon, and in the soft red bark of these are many recesses forming little seats. An occasional small dropping below the edge of some of these suggests that little birds make use of them; if so, it may well be that they are the sleeping or resting-places of tree-creepers.

THE CHAFFINCH

At Fallodon I used to note the 5th of February as a possible date for first hearing the chaffinch; this year I did not hear one till the 13th, but then several began at once. His song is robust and buoyant. Warde Fowler gives a very apt description of it in one of his books. He compares the manner of it to a bowler running with quick steps up to the wicket, and then with an overhand turn of the arm delivering the ball; and he notes that when the chaffinch first begins to sing it cannot for some days 'deliver the ball'; it gets to the wicket and stops. To those who know the song this illustration will surely commend itself. I have known people complain of the persistent iteration of the chaffinch's song, and I must admit that it does suggest a happiness that is a little trivial and commonplace. If the chaffinch were human one can imagine that he would say 'Cheerio!' as a greeting to a friend. It is pleasant to play with these criticisms of a chaffinch: to do so, when one has a sure affection for it, does not really lessen one's feeling for the song, but endears it. The song is so persistent and abundant that from February well on into June it is a common-place in the birds chorus; and yet one chaffinch stands out in my memory as does a single dipper and a single wren. It was in the Whitsuntide recess, when for a few precious days late in May or early in June I had escaped from Parliament and from London, whose 'season' is then so miserably unseasonable. The days were fine and bright. On a stone coping of a little parapet, that went round the roof at Fallodon, at a corner that faced due south, a chaffinch used to take its stand, and from that eminence day after day it pelted me with song whenever I went on the lawn outside. This one bird became a feature of the holiday, an embodiment of happiness proclaimed from the house-top. Wordsworth, in memory of a day spent by 'a little unpretending rill', says:

> The immortal spirit of one happy day
> Lingers beside that rill in vision clear,

and for me the immortal spirit of those happy Whitsuntide days still lingers in the song of the chaffinch.

THE SKYLARK

I came across a small volume of verse written by a young airman in the war. There was no evidence that he knew Wordsworth's poem on the Skylark, but one of the experiences that he described as most pleasing or glorious was the sensation of being high up alone in the sunlight. When Wordsworth wrote 'A privacy of glorious light is thine,' his imagination had penetrated to what man would feel in the skylark's place. Without giving the bird human attributes, he had linked its flight and song to human feeling. This is an achievement in poetry that 'fancy', however beautiful, cannot accomplish.

The skylark becomes an essential feature of country life. I was walking from one plantation to another with an old woodman: it was a fine April day: we were crossing grass fields and skylarks were singing: as bird after bird rose about us the old man said aloud, but to himself rather than to me, 'Them lav'rocks mounts and sings.' They were part of his life. He was an intelligent man, but self-contained; rough in manner with others, aloof and unpopular. He had never seen London, and I arranged for him to spend a week there. We went on the top of an omnibus from Piccadilly to the City. 'Eh! what a throng!' he kept repeating, as he looked down on crowded streets and pavements; and finally, 'It's fair like bees swarming.' He was interested in what he saw in London, and enjoyed his week there, but after his return home, when we were walking alongside an old avenue, he remarked, 'I couldn't bear to live in a city where I couldn't lift up my eyes and see the trees standing one behind another.' If he had been told of the city dweller who after a day or two of the country fled from it because he could not stand the 'tingling silence', it would have produced nothing but the blank stare of inability to understand: so, too, the lover of the town can never understand the sense of depression, exile and exasperation that the country man feels when brought to live in a great city.

THE MEADOW-PIPIT

The meadow-pipit is a bird that, being gifted with only a small song, displays it to the best advantage; it ascends to no great height — perhaps twenty feet or so — and then descends like a little parachute, singing as it comes down. Here again the song is evidently part of a joy-flight, and when one is salmon-fishing in early spring in moorland country, meadow-pipits make a minute but perceptible contribution to the happiness of the day. They are very common in heather country, and though so small and, except when in song flight, so insignificant in their ways, they evidently have a strong scent. To the nose of a pointer or setter a moor must

seem pervaded with the smell of meadow-pipits, and when grouse are scarce a dog will vary the monotony of finding no grouse by pointing a meadow-pipit, or 'tit-lark' as it is often called. The scene is only too well known to every one who shoots grouse over dogs. The pointer becomes as rigid and earnest as if a covey of grouse were in front; the guns, if there are two, draw together and advance one on each side of the dog, which, encouraged by the keeper, moves forward with intense caution. At length a titlark flits unconcernedly out of the heather, the dog relaxes and wags its tail in a deprecating manner, as if asking us to excuse the joke since it is such a small one; and then, as if doubtful how the joke will be taken, starts off again in quest of the scent of grouse. Thus in one way or another the little meadow-pipit mixes itself with memories of Highland sport.

THE GOLD-CREST

It is only when we know the song of the gold-crest that we realise how common the bird is. It likes fir woods, but is not restricted to them; a fir here and there among deciduous trees contents it. One great silver fir in the garden at Fallodon, one cedar on the lawn at Wilsford, is each a whole world to at least one pair of gold-crests. Its voice, as we should expect, is very high-pitched; the little call-notes are like needle-points of sound. The song has little volume, but on a still day can be heard from some distance. It suggests to me a tiny stream trickling and rippling over a small pebbly channel, and at the end going over a miniature cascade. One hesitates to offer such an image of song, or indeed any image of song; some chance suggests one to the mind, and gradually it is so associated with the sound that to oneself it becomes inseparable from it, however fanciful it may have seemed at first: to any one else it may seem so far-fetched as to be inapplicable. It is, however, worth while for those who attend to bird songs to encourage the image that a song may suggest. It becomes pleasantly associated with the sound, and may help to distinguish it. Probably it was this image of the song of the gold-crest that helped me to disentangle it so completely from the song of the tree-creeper.

The gold-crest is not an easy bird to observe, but this is because of its incessant activity, not because of shyness. It is not a shy bird, but this seems due, not so much to tameness as to indifference: the indifference that small insects show towards large things such as human beings. Once when I was sitting among whin-bushes in the New Forest a gold-crest came so close to my face that it suggested the need of reading-spectacles to peruse it.

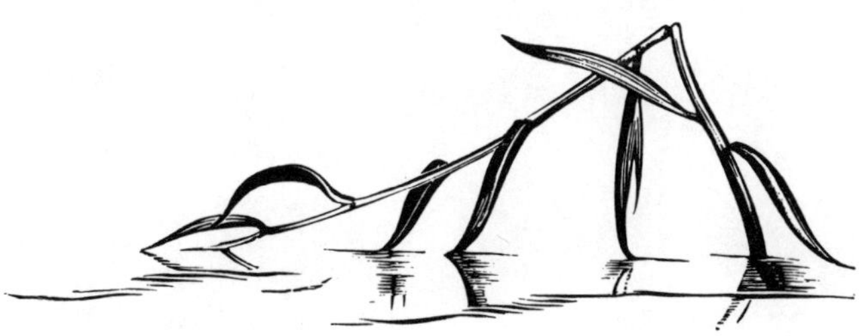

CHAPTER VIII

THE RETURN OF THE WARBLERS

To those who mark the progress of the year by the song of birds, the first hearing of the chiff-chaff is the beginning of a new stage. Hitherto we have noted the coming into song of resident birds; in the chiff-chaff we hear a bird proclaiming not only that he is singing, but that he has arrived. He has accomplished, since we last heard him, two long journeys, one away from us in the autumn, and now the journey of return. Alone of all the warblers, the chiff-chaff has given us the right to expect him in March; he is the forerunner of the rush of song-birds that is on its way to us and will arrive in April, and thereafter enrich our woods, meadows and gardens with still further variety and quality of song. This is why the first hearing of a chiff-chaff moves us so each spring. He is a symbol, a promise, an assurance of what is to come. In himself the chiff-chaff is a very individual and peculiar little bird. I will not attempt to define the word 'warble' or 'warbler'. The term has become so inseparably attached to certain summer birds that there is no choice but to continue the use of it; though if it is to be used at all, there is no justification for its being restricted to the class of birds that are habitually called 'warblers'. There is, however, no conceivable utmost stretch of definition by which the term 'warble' can be made applicable to the sound that the chiff-chaff makes. On the other hand, if anything is to be called 'warbling', the song of the willow-warbler must certainly be put in this category: and the chiff-chaff is so nearly allied to the willow-warbler that they must be thought of as members of one family. In appearance, habits, food, habitat and the manner of nest the two birds have very close resemblance to each other; so close as to make the difference of song remarkable. By the season and manner of utterance the chiff-chaff gives us clearly to understand that he intends his notes to be a song, and as such I, at any rate, accept it from him.

Ill-disposed persons may say that it is nothing more than chirping: against this assertion I should protest. There is a spirit in the two notes of the chiff-chaff that suggests the same motive as song, and there is something more than the mere reiteration of two notes. To one who listens attentively there is the hint of a pattern in the variation of the notes. The sound also suggests industry, as of the passage of a shuttle to and fro. Certainly the chiff-chaff is a very persistent and industrious performer. There is not, as a rule, more than one pair in the garden at Wilsford, but the notes are so frequently uttered for such long periods of the day, that they are one of the most familiar sounds of the garden: if one did not observe that the sound came always from the same region, the impression would be that the garden was full of chiff-chaffs. Like other song-birds, the chiff-chaff becomes comparatively silent when strenuously engaged in feeding the young; but some time in the latter part of June he will for some days 'sing' throughout the day as if he had nothing else to do. The notes then have not quite the vigour of April and May, but they are even more frequently and continuously repeated. Then comes the moult, and the bird falls silent till September, when energy is restored and expresses itself in a subdued repetition of the 'song' — a sort of quiet farewell before the chiff-chaff leaves us on its long journey Southwards.

The chiff-chaff has, in fact, named itself, for the name is suggested by the sound of its notes. The bird has therefore the good fortune to be well named. Its nearest relation, which arrives early in April, is unfortunate in the matter of its name. It is usually spoken of as the 'willow-wren'. The name

is not so offensive and vulgar as the name 'hedge-sparrow' is when applied to the dunnock, but it is inappropriate. The bird is not a wren; this name should be kept for the common wren, a bird so distinct and individual among British birds that its title should be kept for it alone and not shared by any other species; though perhaps a concession to ordinary usage may be made in favour of the gold-crest, which is often called the golden-crested wren. Nor is the so-called 'willow-wren' more partial to willows than other birds that frequent trees and bushes. In places where I have observed willow-wrens to be particularly abundant there are no willows, nor the moist ground that willows desire. It would, however, be futile to invent an altogether new name for this bird: it has already the alternative name of willow-warbler, and I propose to use this name, as being the less inaccurate of the two. Of all our summer warblers this seems to me to be the most widely distributed. I have never been in any part of England where it is not common; in the woods about Fallodon it is so abundant that the song forced itself upon my notice before I had begun to take any account of small birds: all day and every day in May and early June, when I was in the woods, the song of one willow-warbler after another was in my ears: it was impossible to ignore it. Not knowing from what bird it proceeded, I called it 'the everlasting bird', because the songs seemed never to cease. In the South of England one expects to hear it in the first week of April. The song is particularly pleasant as well as frequent. It is a succession of slender and delicate notes, forming a completed sentence, which is repeated again and again at short intervals. The bird is a desultory singer, uttering its song as it flits about the bushes, searching the leaves for insect food. The notes have a very endearing quality of their own. They suggest something plaintive — as if the bird were pleading. 'A cadence soft as summer rain' has occurred to me when listening to the song, which is peculiarly touching. It is entirely without any note of bravado, exultation or challenge, such as is suggested by many other songs. To all who find pleasure in birds, the arrival and first hearing of the willow-warbler each spring is a moving incident. In the summer moult the willow-warbler becomes silent, but in September it sings for a little, infrequently and rather faintly. The resumed song has in it something meditative and reminiscent. This, at any rate, is the mood it suggests to us: then unnoticed, the willow-warbler slips away on its journey to another continent, to winter in Africa.

A bird to be thought of with the chiff-chaff and the willow-warbler is the wood-warbler. It differs from these two species more than they do from each other, but there is considerable resemblance. The wood-warbler (for the reason given already I prefer this title to 'wood-wren') is well named. It is a bird of full-grown woods. It does not frequent gardens and shrubberies, as the chiff-chaff and willow-warbler do: it lives in big trees, especially beech and oak: so indifferent it is to bushes that it seems almost to prefer

woods, where the ground is bare, as is so often the case under beech trees.

The wood-warbler is brighter in colour than the willow-warbler: it looks a stronger bird, and the length of the wings gives the impression of a shortness of tail, that causes the bird to look less slender and more robust.

The song is very remarkable: the bird has in fact two songs, so distinct and unlike that nothing but clear and close view can convince one that the different sounds are made by the same bird. Fortunately it is easy to watch it; for though it is often in the upper part of a big tree, it also descends to lower branches, and there, within a few feet of human yees, it will continue its business of feeding and singing. So occupied is the wood-warbler in these pursuits, so apparently confident in the security of its own joy and innocence, that it seems not to mind being watched. The open branches of big trees do not give the exasperating concealment with which bushes baffle us so often when we are trying to observe other little birds. The time to watch and listen to the wood-warbler is when the beech leaves are fresh and tender on the boughs early in May: the soft green and yellow colours of the bird are in tone with the foliage, and its ways and movements and general happiness animate the beauty of young beech leaves: wonderful and perfect beyond description as this beauty is, the presence of a wood-warbler can still add to it. The song that is most frequent is the shivering or sibilant sound, as it is called in books. To appreciate it one must watch the bird singing, and thus get the impression of delight that is in it. When this has been watched and remembered, the song will always suggest a little ecstasy, whether the bird is seen or not.

The other form of song is one clear and very plangent note, repeated in gentle succession as often as nine or ten times. The bird no doubt expresses the same feelings by this as by the other part of its song, but the impression made on a human listener by these plaintive notes is one of intense sadness: the tone is pathetic; there are tears in the voice. Sometimes a wood-warbler will give these notes at no long intervals, though never so frequently as the shivering notes repeated again and again, at very short intervals, and the bird will withhold the plaintive notes altogether. When this is so, we do not get a complete impression of the wood-warbler, for it is the contrast between the two songs that is characteristic of the bird, and indeed remarkable among all bird songs.

Warde Fowler in one of his books tells how he was puzzled when first making acquaintance with the wood-warbler. He thought he must be in the presence not of a single bird, but of two birds of a different species. The passage was familiar to me before I had heard a wood-warbler, and when I did hear one for the first time I became aware that I was going through precisely the experience that Warde Fowler had described: thus to me the discrepancy between the two sets of notes was not a mystery, but a revelation. I have noticed the wood-warbler to be very abundant on the only two occasions when I have been in the New Forest in late spring, once early in June and once in the latter part of May. It seemed as if I passed, for two or three miles at a time in the heart of the Forest, continously from the song of one wood-warbler to another. At Fallodon there are several pairs in the woods near the house, but the earliest bird is not heard there as a rule till the beginning of May, three weeks later than the willow-warbler.

Two warblers of high rank in song must be considered together — the blackcap and the garden-warbler. There is close resemblance between them in habits, voice and nests: so closely allied are they that one pair seems jealous of the presence of another. Adjoining the cottage in Hampshire was a chalk-pit, overgrown with blackthorn, bramble, wild-rose, and much elder; there were also a few oak, ash, and maple trees. For all the thirty-three years that I knew the chalk-pit intimately, one pair of blackcaps nested there every spring. In the later years a garden-warbler was to be heard singing there early in May, but after some days he would be heard no more, and the song of the blackcap continued alone. My interpretation was that a pair of garden-warblers, attracted by the aptness of the place for their needs, endeavoured to establish themselves in it: the blackcaps, who had been in previous possession regarded the garden-warblers as rivals and prevented them from settling in the chalk-pit. Other warblers nested there: one pair of willow-warblers every season, a pair of sedge-warblers generally, a pair of nightingales, and a pair of each species of whitethroat sometimes, but never more than one pair of the same species. The inference is that one pair of warblers will tolerate another of quite different species in the same territory, but the presence of a pair of blackcaps and the extrusion of the garden-warblers suggest that these two species were too nearly related for one pair of either kind to allow the other to inhabit the same territory. I have not observed the same intolerance between chiff-chaff and willow-warbler, though these kinds resemble each other even more closely in appearance than do the blackcap and garden-warbler. The two latter birds, on the other hand, have a resemblance in song that is rare, if not unique, as far as my knowledge goes, among British warblers. The conclusion is that it is resemblance of song rather than of habits or appearance that provokes animosity between birds of different species.

In appearance blackcaps were very distinct from garden-warblers: the size and shape are alike, but the blackcap is an exception to the rule among our warblers that both sexes should have the same plumage. The prevailing colour of each sex is a soft grey, but the male has a black cap, the female has one of rich brown on the top of the head. The general appearance of both is of elegance and good taste. Garden-warblers have the quiet greens and yellow colours of the willow-warbler — they are, indeed, like larger willow-warblers without an eyestripe, and the sexes are like in colour.

For perfection or moving quality of voice I should place the blackcap with the blackbird and the nightingale in the first class of British songbirds. His song is loud, exceedingly sweet, but also spirited: it is not very long, but is frequently repeated: there is not great variety, but the thing done is absolutely perfect. There is not a note that fails to please or to be a success. The tone does not stir us so inwardly as that of the blackbird, but it is sheer delight to listen to it. Of the blackcap, indeed, it has been said that, like the gipsy before the castle gate, 'he sang so very completely'.

The garden-warbler's song is very good: in one respect it surpasses that of the blackcap; it is more sustained; but the bird never seems absolutely to clear the throat and let out the sound so pure and free as the blackcap does. To my ear the opening notes of the blackcap's song and those of the garden-warblers are so alike that I hear them with a doubt of which bird I am listening to; but the garden-warbler goes on and on for a longer time and yet never seems to liberate its voice upon the air so completely as the blackcap does. In other words, a garden-warbler's song seems always on the point of an achievement, to which only the blackcap attains.

Having made this critical comparisons between these birds, let me end as I began, but putting them together. They are rather larger than any of our warblers except the nightingale, and with this same exception they are greater in song than the others. I have never found either so abundant as the willow-warbler is in some places, but they are very widely distributed; and in all country places where I have lived in England there have been a few pairs every season in the neighbourhood; and though the two birds may not like each other, those who have an ear for bird's songs are grateful to them both.

A number of other birds arrive in April that add to the variety of life and to the animation of English country, but are of minor importance in song. Two that are especially familiar to every dry-fly fisherman in chalk-stream valleys shall be noticed first — the sedge-warbler and the reed-warbler; though very few of the latter arrive before May.

It is impossible to write of the sedge-warbler without using epithets or expressions that apply to mankind rather than to a bird. When such expressions are used here or elsewhere, they are intended, not as literal descriptions of the bird's nature and character, but to give the impression that the bird's appearance and ways produce upon a human observer. The sedge-warbler is like a comic spirit among birds. The eye-stripe gives a jaunty look; its demeanor and movements are very lively. The natural song is apparently akin to that of the reed-warbler, but it is harsh in tone, and the bird interpolates various imitations of notes of other species. Warde Fowler, in a book already quoted, describes how he saw and heard a sedge-warbler imitate, as if in mockery, the fussy chuckle of a blackbird scared by a false alarm and flying off in unreasonable perturbation. The bird will often flit out almost under the angler's foot from some thick growth on the edge of the stream as if its nests were surely there, when as likely as not the nest is in the hedge outside the water-meadow. Sometimes it seems to frequent the reeds; as if it made common cause with reed-warblers; and yet we come upon it in bushes on higher ground in dry places not adjoining water. It is as if the bird wishes to show us that it can be indifferent to the neighbourhood of water, which we have thought indispensable to it. It is lively and noisy all day, but also sings at night. If one sleeps out of doors, various birds, especially the dunnock, may be heard to utter an occasional snatch of song; but with the exception of the nightingale and grasshopper-warbler, the only one of our songbirds that I have heard sing frequently and even continuously at night is the sedge-warbler. The song lacks melody, but the bird is a pleasant and animating presence.

The reed-warbler's name is so literally descriptive of the bird as to be unimaginative and almost dull. It lives entirely in reed beds; it populates large tracts of reeds, such as are found in the Norfolk Broads, but wherever a rood or so of reeds grow we may expect to find a pair or two of these birds. Away from reeds I have never found it; and they must be the common reeds with long thin stems; sedge or other kindred growths will not serve its purposes. It is quiet and soft in colouring, and its song is a monotonous warbling, less varied but more melodious than that of the sedge-warbler, and pleasant to listen to. It sings continuously clinging to the stems of the reeds, among which it is invisible, except at a very short distance. A moderate disturbance will start reed-warblers and also sedge-warblers singing. A friend and I were standing on a plank foot-bridge that led from one meadow to another across the Itchen; one end of the bridge passed through a belt of reeds. My friend was not very familiar with the reed-warblers song, and wished to hear it. A bird was singing in the reeds at a distance of some two or three yards, where we could both listen to it and watch it singing. Presently it ceased, but did not move. 'Would you like to hear more of it?' I asked. 'Yes,' replied my friend. I picked up some small missile and threw it into the reeds in the direction of the bird, which at once broke into song again. My friend was rather shocked, and to this day affirms that I nearly hit and killed the bird.

Reed beds have a delicate beauty of their own: especially is this to be noted on a sunny day in February or March. There is no young green in the reed bed yet, but last year's stems are cream-coloured; the dark feathery heads look silvery in the sunlight and nod like plumes as the wind agitates the innumerable reeds. The warblers have not come, but this is the time when reed beds have a peculiarly delicate and graceful beauty. Imagine such a day on the Norfolk Broads, where the spaces of open water between the cream-coloured reed beds are blue with the reflection of the sky.

As the new reeds grow up green in May the reed-warblers come and make their presence known. The time to listen to them is on a fine day in June, when there is just enough breeze to make a slight rustling in the tall reeds that blends with the continous singing of the birds. Then we know the world in which reed-warblers live, and we feel the spirit of it.

The two whitethroats, the common and the lesser, arrive in April. I do not see much difference in size between them, but the common whitethroat is a little brighter in colour than the other: it is also more common, more conspicuous and lively in its habits, and altogether has a little more of everything noticeable than the lesser whitethroat. Its song is fussy, as if the bird were always in a hurry or slightly provoked. Sometimes the tones and manner suggest scolding. The bird is fond of roadside hedges which have thick undergrowth, and so thrusts itself upon our notice; but it also frequents clumps and woodland, especially where there are patches of bramble. In spite of what is said above about the bird and its song, the prevailing impression it gives is that of excitement and happiness, and its animation and vitality are a pleasant feature in the places that it chooses to inhabit.

The lesser whitethroat, though so like the other in appearance, is different in its habits. It is retiring and unobtrusive. I have not noticed it in roadside hedges but in quiet places, such as the thicket of disused chalk-pit: and it seems to like large trees, to which the commoner bird, though not averse, seems indifferent. It needs bushes for a nesting-place, but does not require a growth of nettles or other coarse herbage as the common whitethroat does. Its song is to me associated with warm days in June or even July: it is a monotonous repetition of one note, and suggests the syllable 'sip' repeated several times; then after a pause repeated in the same way. Altogether it is a curious contract to the other bird in its comparative lack of animation and its quietness.

I have heard the lesser whitethroat utter notes like the song of the common whitethroat, and should have passed it by unless I had also heard the other song, but its characteristic song to me, founded upon observation of a pair that used to nest close to the Hampshire cottage, is the monotonous repetition on hot summer days described above. My observations with the lesser whitethroat are mainly in southern counties. I have, however, heard it more than once at Fallodon, where the other whitethroat is as common as anywhere else.

There are other arrivals in April, noticeably the corn-crake and the cuckoo, but this chapter had best be kept for birds that are, in the usual acceptance of the term, song-birds.

The tree-pipit has a more considerable song than the meadow-pipit, and as the name implies, it is a bird of the trees. From the top of a tree it springs up into the air and descends like a parachute, singing as it comes down to the perch from which it sprung or to another close by. The song is loud, and ends with notes that are very emphatic and determined. Sometimes the bird just sits and sings without leaving its perch at all. Of late years this otiose habit of the tree-pipit seems to have increased. The impression is probably due to accidents of my own personal observation: but the habit is one to be regretted, for the song when part of the little joy-flight, with its beautiful descent in air, is a very distinguished performance in bird song. Indeed, I resent being given the song without the display of flight, to which lovers of tree-pipits feel that they are entitled.

There is one bird classed as a warbler, but so distinct and unlike the others that it is difficult to write of it as a warbler. It comes with the rest in April, and is called the grasshopper-warbler. The sound it makes resembles the whispering noise of a grasshopper, but is more robust and louder: it is also like the noise of a small soft-running reel. In manner, delivery, motive and intention this sound is evidently a song. It ceases therefore after the breeding season, but while it is being uttered the whole attention and energy of the birds are given to it. The sound is made continuously, without apparent drawing of breath, and is kept up for some time. Willow-warblers and wood-warblers may sing intermittently as they flit among the leaves, but when the grasshopper-warbler sets himself to sing he can do nothing else. For this purpose he often chooses some dwarf willow bush in the rough ground that he frequents. There he will sit and 'reel', turning his head occasionally from side to side, a motion which has the effect of making the sound appear now more now less distant to the listener, though the bird is stationary. As the summer advances and nights get warm, the grasshopper-warbler becomes a night singer. At Wilsford there is some rough marshy ground at the edge of the garden near the river. Here in 1922 a grasshopper-warbler sang on a single willow bush, and could be heard far into the night. The bird was not shy when singing, and in daylight could be watched only a very few yards away, so absorbed was it in its own performance. In one mile of the Itchen Valley that had much rough tussocky ground, too rough and coarse to be mown, there used always to be two, and sometimes three, pairs of grasshopper-warblers; and just when the evensong of birds had ceased in the warm dusk of June and July evenings, the grasshopper-warblers would begin to sing. Thus to me the song became associated with failing light and the end of a day's fishing. The territory of each bird became familiar to me, the presence of each was greeted every year and noted evening after evening on the way home in the quiet twilight: my waders, brushing through the lush, soft growth on the river bank, made a sound not out of keeping with that of the bird. Apart from the act of singing, the grasshopper-warbler is not seen: its small dark form is as unobtrusive in its ways as a mouse in the thick rough herbage, where the bird nests. The sound it makes suggests dryness, as if there were no moisture on the palate; the bird endears itself not by melody, but by the unusual and quaint character of its performance; its manner is very quiet, but the length of time for which the song is sustained gives an impression that the bird takes exceeding pleasure in it.

CHAPTER IX

LATER SONG

THE CUCKOO

The cuckoo makes no nest: it dumps its egg, the hatching and the care of its young on some pair of birds of another species, whose domestic affairs are thereby spoilt. The cuckoo has no family life, and is said to be polyandrous. All this has aroused the disapproval of mankind. I do not know how it has been proved that our cuckoos are polyandrous; the males certainly seem to be much more numerous than the females; but it has to be remembered that the familiar notes of the male correspond to the song of male birds; these notes would therefore be heard much more in the breeding season than the notes of the female, even if the two sexes were equal in number. The female cuckoo has notes in the breeding season that are as distinctive as those of the male among the notes of other birds; but these female notes are much more seldom heard. In fact, to many even of the people who live in the country , these notes are not a familiar sound, and, when heard, are not known to be those of a cuckoo. They are sometimes called the 'water-bubbling' notes; and I cannot improve on this description. They seem to me to be used solely as a mating call. I once had a particularly good example of the female cuckoo's call. I was sitting under a lime tree at the Hampshire cottage; the spreading branches concealed me from outside view, but looking out from beneath them much could be seen. A cuckoo came flying and lit on the branch of a poplar only a few yards in front. Having chosen its perch, the bird uttered the loud 'water-bubbling' notes.

A male cuckoo was immediately heard at some distance; it came, cuckooing vehemently as it flew, and lit on the branch by the side of the first bird. There what Chaucer calls a 'spring observance' took place. In this instance there was no sign of a second male in the neighbourhood; there was nothing, therefore, to suggest that the two cuckoos were not paired like monogamous birds. Sometimes, however, the 'water-bubbling' note provokes the cuckooing and pursuit of more than one male bird. The familiar 'cuckoo' is thus, on some occasions, an answer to the call of the female; but generally, like other birds' song, it is an expression of sheer joy, vigour, challenge or excitement.

The voice of the cuckoo, apart from its habits, has always been popular. The old English verse, 'Summer is icumen in', is a warm welcome to the bird. The instance where Wordsworth transfers the melancholy of his own mood to the cry of the cuckoo, is an exception. Elsewhere the cuckoo's note is a cause of rejoicing, a suggestion of happiness.

Every year it is welcome because it is so intimately associated with the most beautiful season of the year. It is a strange cry that does not blend with other sounds of Nature, but interrupts them or breaks a silence, like a human voice.

> Breaking the silence of the seas
> Among the furthest Hebrides.

I have never been in the Hebrides in cuckoo-time, but I have found cuckoos particularly abundant in June in the glens of the Western Highlands of Scotland. As the season advances the bird plays various tricks with its voice; the straightforward clear 'two-fold shout' may be broken up into imperfect syllables. The cuckoo also makes an odd noise, a gruff chuckling sound, that has been described as goblin laughter.

THE REDSTART

The redstart is not a very important singer, but is sufficiently so for its arrival to be listened for every year and for the song to be picked out by ear and attended to as we walk about the woods. For some reason the bird does not seem to care for the chalk-stream valleys: neither about the cottage on the Itchen nor at Wilsford in the Avon Valley have I heard it. At Fallodon there are three or four pairs every season. I called a friend's attention to it there this year: he did not know the song, and after listening for a while said that the first part of the song reminded him of the chaffinch. At that moment a chaffinch sang close to us, and we listened to the two birds: the contrast between the robustness of the voice of the chaffinch and the slender notes of the redstart destroyed all resemblance. The redstart's is a thin song, but there is something brisk and wild in its tone that gives distinction. The first notes seem to be uttered with ease; the last notes sound as if they caused the bird a little effort and uncertainty. I have once heard the song as far north as Sutherland before the end of April.

There is one aspect of the redstart that differs from that of all these April-coming songbirds. None of these are brilliant in plumage; their beauty is quiet rather than gay: the plumage of the male, with the exception of the blackcap, is the same as that of the female, and does not alter much, if at all, after the breeding season is over. But the cock redstart has colours that arrest the eye: he is indeed one of the most beautiful of our birds; and it is a plumage of the breeding season; afterwards comes an eclipse that places him more nearly on a level with the female; but in all redstarts the patch of reddish colour on the tail coverts, from which they get their name, is constant.

THE NIGHTINGALE

Two friends, one English and one American, each with knowledge and love of birds, were to spend two days with me in Hampshire in the latter part of May. To hear birds in song was the object of the visit. In the previous weekend I stood near my cottage and listened for nightingales. Time had been when, for some years, a pair had bred in the old chalk-pit close to me; but that time was no more. One year the male had been in full song day after day and night after night; then one night in mid-May there was a silence that was never broken there by a nightingale again: a stealthy and deadly cat, at which I never could get a shot, frequented the thickets of the chalk-pit that year; the nightingale, and also a tame thrush that had trusted me for four years and come confidently into the rooms, both disappeared. No other thrush ever took its place, and no other nightingale came to the chalk-pit. So now I had to listen afield. I heard two nightingales at some distance, one close to the village about a quarter of a mile away; the other even more distant. A week later my friends arrived. Nothing was to be heard near the village, and about ten o'clock in the evening we went to the more distant place; but no nightingales were to be heard there either. A sedge-warbler offered himself as substitute, or, it may be, simply mocked us, and we retired unsatisfied to the village inn, where, in default of sufficient room in the cottage, my friends were to spend the night. There I left them, and to prevent future disappointment went in search of another nightingale in some place more remote from village cats. One bird was heard, and the place carefully

noted, in a large wild park with oaks and thorns and brambles. The next night we went to the place; the bird sang; we crept close to it; no burglars fearing detection could have approached more cautiously or whispered more low.

The moon was rising as we crouched and listened: the bird's song was the only sign of life: for some time we remained; the bird sang, and paused and sang; finally our American friend said, 'Let us go away quietly and leave it singing.' We withdrew as cautiously as we had approached, and the bird sang on. It was no better than many another nightingale, but once again there had been a moment that lives in memory. The trouble taken to find the bird one night, the satisfaction of finding it the next night, the success in getting close to it without disturbing it, the rising moon, the stillness of the night, the apparent solitude of the bird, the whole circumstances and setting, all combined to make this one song memorable.

The conventional nightingale of literature is taken from ancient legend: it postulates a doleful bird, a female, perpetually lamenting, even leaning its breast against a thorn to make its mourning more poignant; this has no more resemblance to the real bird than heraldic emblems have to real creatures. Not so does Wordsworth write of the nightingale:

> O Nightingale! thou surely art
> A creature of a fiery heart: —
> These notes of thine — they pierce and pierce;
> Tumultuous harmony and fierce!
> Thou sing'st as if the God of wine
> Had helped thee to a Valentine;
> A song in mockery and despite
> Of shades, and dews, and silent night;
> And steady bliss, and all the loves
> Now sleeping in these peaceful groves.

That is the authentic nightingale; it explains too why, if the nightingale's song is the most marvellous, it is not the most loved of English bird songs.

The nightingale's song has compass, variety and

astonishing power; it arrests attention and compels admiration; it has onset and impact; but it is fitful, broken and restless: it is a song to listen to, but not to live with. Wordsworth, in the poem already quoted, explains this feeling; he goes on to describe and to express his preference for the cooing of the wood-pigeon to the song of the nightingale. Here are some of the lines.

> He did not cease; but cooed — and cooed;
> And somewhat pensively he wooed:
> He sang of love, with quiet blending,
> Slow to begin, and never ending;
> Of serious faith, and inward glee;
> That was the song — the song for me!

THE YELLOW-HAMMER

The yellow-hammer, our gayest bunting, begins his song earlier than all his common British relatives. The formula for his song, 'A little bit of bread and no cheese', is well known. The yellow-hammer, like the chaffinch, for the first few days has difficulty in finishing the phrase. He will accomplish the first part of it, and then stop. In fact, for some days he may be said not to be sure about his 'cheese'. The yellow-hammer is a great frequenter of roadside hedges, on the top of which he will sit and sing, and is one of the prettiest and most familiar sights of country walks. In the last week of February this year I found myself on such a walk passing between two yellow-hammers, each singing perfectly in alternate rivalry.

The yellow-hammer's song is continued very late in the summer, even into August. It is not a song of power or melody; its merit is in quaintness and familiarity: it draws attention to the bird when he is perched conspicuous on the top of a hedge or whin-bush, and it is always worth while to observe through field-glasses the uncommon yellow head and the exceedingly rich brown colours of the back, for the mature cock yellow-hammer is a very handsome bird.

THE CIRL-BUNTING

In July, when so many songs have ceased, it is worth while to note the birds that still continue in full song. Larks, wrens and yellow-hammers have been familiar to us in the earlier months, and we hear them still; but there are some minor songs to which, though they began weeks ago, we now have time to give special attention. Three of these are buntings, that come into song later than the yellow-hammer. The cirl-bunting is the best singer of the three; it is a bird that is common only in certain districts. The song resembles the first part of that of the yellow-hammer, but is fuller in tone; it has, however, not the variety of the yellow-hammer's song; it suggests to me the ringing of a small electric bell of pleasant tone and good quality of sound, but monotonous. My first hearing of it was an incident not to be forgotten. I was bicycling back from the Test to the Itchen Valley; I had taken field-glasses, for the object of the ride was to look out for birds. I left the Test Valley at Stockbridge and had reached the turn beyond Lainston, where a side road makes a short cut to the left to Harestock: at the turn I missed the field-glasses and knew that I must have left them at the inn at Stockbridge, some miles distant: the road has many ups and downs, and it is no pleasant matter to retrace miles just travelled with the knowledge that after retracing them one must come back over them again. The consciousness that nothing but one's own carelessness is to blame makes the process exasperating. The impulse is to take it out of oneself by exercise, as if punishment of the body were some compensation for forgetfulness of the mind. One is eager to get back to the starting-point as soon as possible, and thus obliterate the loss of time and temper caused by the affair. There was no dismount till Stockbridge was reached: there I laid hand on the glasses, remounted immediately, and took the hill on the Winchester road out of Stockbridge without getting off. No mean performance for a member of Parliament in the middle of the sedentary session in days when there were no change-gear bicycles. At the very spot where I had missed the glasses and had resolved that I must go back for them, I heard a song new to me. In a moment there was a dismount, and the glasses were out of their case; the bird was sitting on the top of one of the bushy yew trees that are so common in hedges in chalk country, and I was looking at and hearing a cirl-bunting; pictures of the bird were well known to me, and the lemon yellow and chocolate colour about the head made it unmistakable. This was more than compensation for time lost: indeed, hearing the song of the bird at that very spot seemed like a reward for the resolution taken there, though the spirit of wrath in which I had ridden those extra miles to Stockbridge and back had merited no such gracious occurrence. Afterwards I came to know the bird as common in that district, and W. H. Hudson so describes it in *Hampshire Days*.

THE REED-BUNTING

The cock reed-bunting is very handsome, and the birds are common on ground that suits them, such as chalk-strewn water-meadows and reed and sedge growths. At Wilsford, where water-meadows adjoin the garden with no fence to separate them, the reed-bunting is a garden bird. In spite of his noble appearance the reed-bunting has rather a paltry song: it suggests to me the ascent of steps, the first two or three being mounted sedately and the last taken trippingly. In the plenitude of song in May and and early June the reed-bunting's contribution would not be missed, if he happened to be absent or silent, but in July by a chalk-stream I should miss him very much; there he prolongs the season of song. The regard felt for his song is an example of how valuable the insignificant may become when our attention is given to it, and it is a part of pleasant associations. The reed-bunting is very conspicuous, and places himself well for observation: when his song has given notice of his presence, it is worth while to look at him through field-glasses.

THE CORN-BUNTING

It is not altogether kind to write much or tell the whole truth about the corn-bunting — his person or his song. The tendency of buntings is to be robust rather than slender, but the corn-bunting is the cart-horse among them all. he has the habit, too, of sometimes taking a short flight with legs hanging down, as if it were too much trouble to tuck them up neatly in flight like other birds: this adds an impression of slovenly disposition to clumsiness of body. Though the largest of our four common buntings, this bird is the least beautiful in plumage. The males of the other kinds all have some distinction of vividness of colour. The male corn-bunting is content to be as dull as his mate, who in turn is duller than other females of this tribe. For a perch the bird prefers telegraph wires or a wire fence. Corn-buntings must have been pleased when the abomination of wire took the place of green hedges or honest posts and rails grown and made on the country estate. I can even imagine that corn-buntings *like* barbed wire. On wire then, by preference, the corn-bunting will perch and grind out the noise that is his song. This suggests to me the sound of two hard pebbles triturated one against the other. There is neither melody nor pleasant pattern in his song. On the Hampshire and Wiltshire downs, where there are plenty of telegraph wires and where wire fences are sadly on the increase, it is impossible in mid-summer to miss the corn-bunting and his 'song'. he thrusts himself upon our notice. I cannot think of one point in which the corn-bunting is not inferior. Even the egg is less remarkable than are those of buntings generally. Yet at last, in spite of the inferiority of the corn-bunting, of which the bird itself is so completely unconscious — for its whole demeanour is that of self-satisfied content — one gets a sort of humorous affection for it.

THE BRAMBLE-FINCH

Bramble-finches are due to arrive in October: they appear in small parties or flocks. If it is a good beech-mast year, they may be seen in great numbers under beech trees. At Fallodon the beech-mast is a very uncertain crop. In some years the sharp three-cornered husks are nearly all empty: in other years they are plump with kernel. In a good year I have seen the road under overhanging beech trees whitened by kernels that the traffic had crushed. In one such year, when a hard winter came, there were immense flocks of bramble-finches busily feeding under the beech trees. In mid-winter I calculated that there were far more bramble-finches than there were birds of any other single species on the place. They were very conspicuous on the thin covering of snow under the trees.

We understand from coloured plates in bird books that the bramble-finch in the nesting season has a black head and neck. I have only seen it in winter plumage, when a careless glance might not distinguish a party of bramble-finches from chaffinches: a field-glass would show the difference. There is, however, no difficulty in identifying bramble-finches as they rise in flight. When the wings open there is the appearance of a white stripe between them, which is very evident and destroys at once, even to the naked eye and at some distance, the resemblance to a chaffinch. The chirp, too, is distinct. Someone who was asked to describe a certain sound gave the bewildering answer that it was 'a noise like a nut'. To me the note of the bramble-finch has always been a 'nutty' chirp.

In the winter referred to above, the bramble-finches came into a wire enclosure where the waterfowl were fed with wheat. I caught a cockbird, and it seemed to me that with a pair of scissors, by carefully cutting off the fringe of each feather on the head and neck, the bird could be turned into breeding plumage. It may be that this change is brought about merely by natural fraying of the edges of the feathers;

but this must be referred to those who have more scientific knowledge of changes of plumage.

THE GREAT GREY SHRIKE

Another winter bird, still more rare, I have seen only once. It was on a January afternoon, a Sunday; I was walking along a plain public road on my way to post letters: there being no postman on Sunday, we often had to do this, and it was looked upon as a somewhat humdrum business, for the post-office was three miles away and the road commonplace. Suddenly my attention was aroused by the note of a blackbird; it was more than the ordinary alarm note which blackbirds give when disturbed: it was a note that suggested terror. The blackbird flew along the hedge and sought cover therein; after it came a bird of about the same size, with smooth flight and rather a long tail: this second bird lit in a small bare hedgerow tree in front of me, and gave me a clear and satisfactory view of a great grey shrike. It did not appear to me that the shrike was actually in pursuit of or seriously threatening the blackbird, and the terror of the latter at the mere appearance of the shrike was remarkable. A year or two later, when I was in London and could not get home, it was reported to me early in March that a great grey shrike was seen for some days in a plantation near the house. It frequented a spot where long-tailed tits were used to build every spring. One of these tits was found slaughtered while the shrike was about the place, and there was no long-tailed tit's nest in this plantation that year.

THE NIGHTJAR

The 'churring' of the nightjar [is] a most soothing sound, continued for long periods without a break. No one unacquainted with it and hearing it for the first time would guess that the noise was made by a bird at all. It is of that class of stationary, soothing, continuous sounds, such as the hum of a threshing machine, or the noise of waves on the shore heard at a distance, which dispose us to sit still and listen indefinitely.

I have not seen much of nightjars, but I had one curious experience, though not of a joy-flight or sound. It was evening, early in September; I had been sitting for some time at the foot of an ash tree that stood solitary in the middle of broad water-meadows. I noticed a bird fly from the branches above me, take something on or near the grass, and return to the tree. This action it repeated several times, and it took little flights about the tree. It was a nightjar, and I enjoyed the opportunity of watching it; for the bird was flying all about me and yet was unconscious of my presence. The bird was silent till at length in one of its flights it passed quite close to and saw me. So near was it that I could see it see me. It gave a piercing shriek, such as I had never heard before from any bird, and flew straight away out of the meadow. For a moment when it discovered me its head had turned in my direction, and the shriek seemed to be uttered *at* me. It suggested not so much fear as rage and loathing: as if the bird was suddenly aware that, unknown to it, a human eye had been watching it, when it believed there was security and privacy. The wife of Candaules, when she knew that the eye of a man not her husband had covertly seen her disrobing can hardly have felt more horror and indignation than was expressed by that nightjar's shriek.

THE CURLEW

Of all bird songs or sounds known to me there is none that I would prefer to the spring notes of the curlew. I have seen the bird finish its notes on the ground after alighting, but I have not observed if it ever gives them without any flight. As a rule the wonderful notes are uttered on the wing, and are the accompaniment of a graceful flight that has motions of evident pleasure. The notes do not sound passionate: they suggest peace, rest, healing, joy, an assurance of happiness past, present and to come. To listen to curlews on a bright, clear April day, with the fullness of spring still in anticipation, is one of the best experiences that a lover of birds can have. On a still day one can almost feel the air vibrating with the blessed sound. There is no rarity about it where curlews breed: it is to be heard through long days in April, May and far into June. In autumn and winter curlews resort to estuaries and the seashore, and the call note is melancholy; but even at this season on a mild day one may be surprised to hear a single bird give a few of the joy-notes, just enough to revive memory of the past spring and to stir anticipation of the next one.

THE SNIPE

Another joy-sound is the so-called 'drumming' of the snipe. It is said to be made by the outside tail feathers. I have not verified this statement, but I do not doubt it. The sound is made during a prolonged joy flight. The bird flies round and round over a wide area at some height, and every now and then descends in the air at an angle and with accelerated speed; it is in the descent that the sound is made: indeed, this seems to be the whole object of the descending motion, for the bird then immediately recovers itself in the air and resumes its flight. The whole proceeding is one to be watch-

ed. The flight has a very happy appearance, but the 'drumming' seems to be performed by the bird in order to find vent for an exuberance of spirits that cannot be expressed by flight alone. The sound is that of a bleat rather than of a drum, and in some places the snipe is called 'air-goat' in consequence. The snipe has another note that I have heard only at the same season as the 'drumming.' This is a very monotonous 'chucking' noise. It may be uttered on the wing, but the bird takes pleasure in doing it on the ground. When I was lunching out on the Broads one day in late spring, a snipe stood in view on a heap of cut reeds and made this 'chucking' noise till it became wearisome. This snipe did nothing else, but I have heard the noise made on the ground in a water-meadow, and have presently seen a snipe, presumably the same bird, rise up from the spot and do its joy flight and drum. Whether the female snipe makes this tiresome note as well as the male, I do not know. It has no resemblance to the ordinary note uttered at all seasons when the bird rises from the ground.

THE BROWN OWL

Of bird sounds still to be mentioned is the noble hoot of the brown owl. It is not peculiar to autumn, but it is very noticeable in autumn and in winter, as Shakespeare knew very well:

> When icicles hang by the wall
> Then nightly sings the staring owl,
> Tu-whit;
> Tu-who, a merry note.

Poets invest the hoot of the owl with whatever feeling suits their mood or the context of the poem. Gray, thinking of it in the melancholy mood of evening by a churchyard, says:

> The moping owl doth to the moon complain.

And the hooting has often been regarded as bodeful. The normal hoot consists of one fine, smooth long note, then a pause of about four seconds; then a long note, quavering at first and ending smooth and full. Sometimes the bird will give the first long note without following it up by the quavering note, but this is a broken and imperfect performance. I have occasionally, when standing unperceived near an owl, heard it go on for some time uttering one plain but fine-toned hoot with very little pause. The hoot of the brown owl enriches our woodlands, and without it they would be seriously impoverished. If the owls have a period of silence when they may be expected to hoot, I become restless and anxious lest they should have left the place. The hoot is, I suppose, the note peculiar to the male bird when he is in vigour; it is often answered by another bird at some distance, and appears to be like some bird songs, a proclamation of territory. A note which seems to be simply a call-note sounds like 'ke-wick'; in the time when young owls are about it is sometimes repeated in a single tree, with such frequency as to be tiresome, if near the house.

CHAPTER X

NESTING AND FAMILY LIFE

THE CHAFFINCH'S NEST

Every chaffinch's nest is a delight on account of its neatness: indeed each one seems neater than our recollection of previous nests. As described in an earlier chapter, there is a touch of commonplace about the bird and its ways, but there is real distinction and refinement in the nest: it surpasses all other nests that I know in neatness. I have never seen the cock chaffinch take any part in building the nests or in sitting on the eggs. One interesting experience of the resourcefulness of a hen chaffinch came under my notice. The bird had nested in the creepers on the cottage: it had been sitting for a few days on three eggs; one morning the nest was empty: some animal or bird had robbed it. Later in the day a hen chaffinch was flying industriously to and fro between the cottage and a place in a hedge some twenty yards away. The repeated flights invited closer attention; the bird was engaged in removing the empty nest piecemeal and reconstructing it in the hedge. Business in London prevented me from noting how many days were occupied in completing the nest in the hedge, or whether any new materials were used for it: but all the material of the nest on the cottage was removed, and the bird again sat on three eggs in the reconstructed nest: it is curious that the clutch should have been so small in the first as well as in the second laying. Unhappily the eggs were again taken by some marauder, and this time the nest remained empty and abandoned. There was no sign by which to identify the mate of this hen chaffinch: he took no part in these proceedings. He was presumably one of two or three cock chaffinches that were within hearing of the place, and probably continued to sing with undiminished spirit.

THE KINGFISHER'S NEST

One particular nest of a kingfisher may be described for the sake of its attractive secrecy. A small side stream of the Itchen ran for some distance entirely concealed by trees at the edge of beautiful pleasure grounds. Yews, chestnuts and other large trees stood thick together on one side of it; on the other side were willows; the canopy of boughs hid the stream, and it was known to us as 'the hidden brook'. A willow tree was blown down and lay prone on the water-meadow: a great block of bank was lifted and held upright by the roots of the tree: so high was this and so far over did the boughs spread from trees on the other side, that the gap left by the fallen willow did not expose the stream or disturb its privacy. Into this lifted bank kingfishers had tunnelled from the inner side. By going in among the trees the hole could be seen, but could not be reached even on its own side without wading in the water. Here a pair of kingfishers nested every year, flying up and down the green tunnel in which the stream ran, and going to and from their nest unseen. In course of time the earth began to wear away. On the outer side between two of

what had been surface roots there came to be a little opening. By putting an eye to this the eggs could be seen; and by inserting a small twig I once extracted an egg. I was thus in possession of a kingfisher's egg that had been obtained without disturbing the nest — a rare privilege, for, as a rule, I suppose a kingfisher's egg can only be obtained by digging out the nest. The kingfisher's eggs are called white: correctly, no doubt, but they have a delicate tinge of colour, due to the yolk showing through the white shell, which is unusually transparent. I do not remember that there was any horrible smell about this nest, but the dark slimy-looking stuff that oozed out of the entrance used by the birds suggested foulness within the passage, if not in the nest itself.

For a few years after this nest was first found, it was matter of importance to be sure that the birds nested successfully there each season; but gradually the weather and the natural processes of decay wore away the earth, till it became a mere skeleton of decaying roots, and no one who had not seen it as it had been could have imagined that it could ever have been a kingfisher's breeding-place.

THE DABCHICK'S NEST

To say that nests may be underground, on the ground, in or on houses, in bushes or in trees, is a fairly comprehensive statement; but it is not complete. Some nests of grebes may be said to be in the water. The little grebe or dabchick's nest is a blob of dark, dank, sodden weed: it is often placed in a bed of weed in the middle of chalk-stream, where the bird when sitting is in full view. But the sitting bird also has a clear

view, and whenever a suspicious object approaches the grebe covers its eggs and goes off, so that the nest appears only as an uninteresting patch of dead weed. This ruse of the dabchick is a sight very familiar to chalk-stream anglers; as the angler crosses the fence to enter a water-meadow he may see a dabchick rise up on a patch of weeds in the river, hurriedly pull weed over its eggs, then dive with the least possible movement, and so slip away unseen to some concealment. The bird will do this when the angler is still 50 or 100 yards away. The habit of most birds is to sit motionless and tight on a nest; the nearer it is approached the closer will the bird sit: it will not move till in imminent danger of capture or even till actually evicted by touch. This is an example of endeavour to escape notice by the concealment of stillness. The dabchick is an example of the alternative method; that of leaving before a possible enemy has come near enough to see that there is a nest. I have known a spotted flycatcher practise this method when sitting on eggs. One that used to nest on the Hampshire cottage would fly off the nest in a light unconcerned manner whenever a human head appeared round the corner. Chaffinches, on the other hand, would sit very close, and if put off the nest would leave it with fluster and distress.

That dabchicks can fly is proved by their temporary appearance at migratory seasons of the year in unwonted places, to which they can only have come by flying; but diving is their usual method of movement. When the object is to escape, the dabchick dives with the least possible mark on the surface of the water. I have had one singular experience to the contrary. I was kneeling on the bank of the Itchen and putting my fly to a rising trout. My knees were at the water's edge: I bent forward for the purpose of casting: my attention was concentrated on the spot where the trout was rising. I became aware that something was going on in the river close to me and that I was being sprinkled with drops of water. I looked down to discover what unusual thing was happening, saw a dabchick come to the surface close to me, and dive with as much splash and disturbance as possible, throwing up, as it did so, some drops of water that reached me; the bird repeated this exploit several times. In front of me in the thin strip of sedge that grew in the water was the dark blob of the nest; the unwonted conduct of the bird was a desperate attempt, if not to drown, at least to annoy and so get rid of me. I have read somewhere that a flock of coots, when threatened by a hawk, will all dive together, throwing up as much water as they can when the hawk stoops, as if in the hope of drowning it. No doubt my dabchick had the encouragement, when it returned to its nest, of thinking that its little efforts had saved the eggs from an impending monster.

One pretty way that the dabchick has with its young was shown me on the water in St James's Park. When I was first in office and kept in London from 1892-95, I made acquaintance with the man who then looked after the waterfowl, and who lived across the isthmus that is opposite the windows of the Foreign Office. At his cottage I used to call in the season to hear news of the breeding waterfowl, and he would show me various nests. One morning as he was taking me round the island he pointed out a dabchick's nest attached to some willow branches that hung into the water. When we came near he exclaimed that the eggs must have been hatched since he had seen the nest earlier in the morning, for the nest was now empty. We heard a curious little noise on the water, and looking out beyond the branches saw the parent dabchick and her lately-hatched young ones. Being suspicious of us, she had warned the young and now presented her body to them as they sat in the water. Instinct told them what was required: each bird got on to the back of the old one, and was there covered by her folded wings. When all the young had mounted, the parent swam away with her whole family, compact, concealed and safe.

NEST OF THE LONG-TAILED TIT

If I had to give a prize for nest-building amongst our common birds I would give it to the long-tailed tit, which is a bird distributed over the whole of Great Britain. You meet with it frequently, from Sutherland to the South of England, in every county where I have been and where there is anything like a reasonable amount of plantation. It builds a most elaborate nest, and the whole time taken for the building of the nest, the hatching of the eggs, and the fledging of the young is a very long one — much longer than in the case of any other British bird I know.

When I was in office I had a cottage in the country to which I went at weekends, and one Sunday morning before the middle of March I observed from the window a pair of long-tailed tits building their nest in a sweet-briar hedge. When I went out and looked at the nest it was then like an ordinary nest, cupshaped. A long-tailed tit is not content with that, but it builds a nest like a bag with a hole near the top. Every week I went down there the building and business of the nest was going on. It so happened that this particular day, about noon, was the time when the young birds first came out of the nest. It also happened that I was standing close by the nest at the time when the little birds first came out of it. Thus you will see that this pair of long-tailed tits required about two months and a half from the time they began to build their nest to the time the young came out of it. During all those weeks, when I could be there, the nest was a subject of interest to me, and many of you living in the country may have the same experience provided that you will yourselves, and are able to induce other people to observe the rule, not to disturb or destroy.

Long-tailed tits are particularly interesting from another point of view. The birds go in a company and the brood remains together all the autumn and winter, but early in March you will see the long-tailed tits in pairs, and if you will look closely you will see about that time that they have some little nesting material in their beaks, and if you watch them you will see them going to the nest and you can locate it. Sometimes it is rather high up in the fork of a tree, generally oak or ash, but as often as not it is only four or five feet from the ground in a gorse bush or an ordinary hedge. Whether they think they are so small that you do not seek them or whether they are so intensely busy in their work, it so happens that they are not at all shy, and you can stand at a distance of three yards from the nest quite openly and watch them build. You will see first one bird and then the other get into the nest when it is in cup shape and make it round and smooth by rubbing its breast round and round against the wall of the nest. You will see it arch its head over the sides of the nest and pull some of the outside over and inwards, weaving it thus; and when the nest is completed outside, the birds will line the inside with feathers. Then if you like you can help in the building of the nest. If you can collect small feathers and put them close to the nest, you can stand near and see the long-tailed tits take the feathers that you put for them and use them in their nest-building. When the young birds have fledged you can, without doing any harm, take the nest and examine it, because I have never known long-tailed tits use their nest a second time. I have been told that the feathers used in the lining of a long-tailed tit's nest have been counted to number more than nine hundred.

THE CIRL-BUNTING'S NEST

The cirl-bunting ... is a very pertinacious breeder if the following instance is a guide to the habits of the species.

It was the early morning of 27 June, 1900. I was shaving in the little dressing-room at the cottage when I became aware of a cock cirl-buting sitting on a branch of a young acacia a few yards off in front of the window. His manner was uneasy and embarrassed. He evidently saw me and did not like me. If he objected to my presence he was free to go away: why, then, did he remain? The answer was clear enough: he wanted to do something, and he did not want me to see him doing it. The thing that birds most dislike is to be seen going to a nest: from the behaviour of the cirl-bunting it was evident that he had a nest and that it was near to me. I had not hitherto seen a cirl-bunting's nest: no doubt this one could be found now that I knew whereabout it must be; but it would be convenient if the bird would show me the exact place. To disarm suspicion I pretended not to see him and went on shaving. Before I had finished he had flown into a place in the thuja hedge about five yards from my window, remained there for some seconds, and then had flown out again. My note made at the time does not mention that I saw any food in his beak, but probably he was feeding young birds. I hurried to the spot and found the nest, with two very young birds and one addled egg. Till 5 July all went well with the nest: between that date and the 7th some villain, probably a jackdaw, took the young birds. We were not at the cottage after 10 July, and the futher efforts of this pair of cirl-buntings were related to us by W. H. Hudson, to whom the cottage was lent for the rest of that summer. The birds made another nest in the sweetbriar hedge at the other side of the cottage, and this time successfully reared three young birds. Having accomplished this, the indefatigable pair made another nest in another part of the sweetbriar hedge: eggs were laid in this, but it was deserted: probably the hen bird felt that for one season she had done enough. After this there were high hopes that this little plot of home ground would be honoured with a cirl-bunting's nest every season; but it was not to be. There was no cirl-bunting's nest there in any subsequent year. Cirl-buntings, which had been plentiful in that region, became scarce; and for some years before my cottage times came to an end in 1922 I had not heard a cirl-bunting within earshot of the place. It is often so with anything specially interesting observed in nature: hopes are raised and then disappointed, or it may be the reverse: there is increase and decrease, ebb and flow, waxing and waning.

A BLACKCAP'S NEST

The slackening of song while the birds are hard-worked in rearing the young does not need explanation: when the young require no more care the old birds are liberated: their energy and vitality are still great, and till these are repressed by the moult, the male has little else to do than to indulge the impulse to sing.

One observation on the chalk-pit blackcaps will illustrate the point. The nest was found when it was still only a beginning; the male helped in the building of it, and this work caused no great diminution of his song: the nest was in a small thicket of wild-rose, and the cock bird, after having made his little contribution to the nest, would sometimes sing in the rose bush before he left in search of more material. He took his share of sitting on the eggs: he was not observed to sing while actually himself sitting on the nest, but his song was pleasantly frequent till the young were hatched. Then he became almost silent. Early one morning from my bedroom — it must have been about a week after the young were hatched — I heard the blackcap singing repeatedly. I guessed with foreboding what had happened and went to the nest. It was empty: the young had been safe, but not ready to fledge, the day before; and now in the early morning something hostile, probably jackdaws, had taken them. How the female blackcap was affected there was no evidence to show, but the male gave the impression of joyous liberation and high spirits.

THE CUCKOO AND OTHER BIRD'S NESTS

Once in late May or early June, while driving along a country lane, a cuckoo flew out of the hedge with much perturbation. We stopped to examine the spot, and found there a nest not quite finished, but sufficiently advanced to be identified as that of a dunnock. I was not to be in the neighbourhood again for some time: I asked my companion, who was resident, to ascertain what happened to the nest. He informed me afterwards that it had been completed and a cuckoo's egg had been laid with those of the dunnock. This suggests that cuckoos mark down nests in advance.

The experience of a pair of pied wagtails that nested every year on the Hampshire cottage is an example of the persecution inflicted by a cuckoo. For some years the wagtails were unmolested; then came a year when a cuckoo discovered them. Their nest was in a tangled mass of creeper that filled a small alcove on the front of the cottage. The cuckoo placed an egg in the nest and the wagtails hatched the egg and reared a young cuckoo, to the exclusion of their own offspring. When this task was completed, the wagtails built a second nest in ivy on the back of the cottage. Again the cuckoo put an egg in the nest, and the unfortunate wagtails reared a second monster. The next year the wagtails made a nest once more in the same creeper-filled alcove. After three or four eggs had been laid I found a cuckoo at the nest: no egg had then been laid by the cuckoo, but the wagtails deserted and never built on the cottage again.

Cuckoo's eggs found in nests on or close to the cottage were all of the same type, resembling the eggs of pied wagtails; but they were found in various nests; robins and sedge-warblers and dunnocks were used as foster-parents, as well as pied wagtails. There was one exceptional experience with a sedge-warbler. I disturbed the bird on its nest, and a casual glance showed that there was only one egg: I withdrew immediately, fearing that I had disturbed the bird in the act of laying and that it might desert. On returning at the next weekend, I found to my satisfaction that the sedge-warbler was sitting. I put the bird off, expecting to see a full clutch of sedge-warbler eggs; but there was still only one egg, and a more careful look showed it to be that of a cuckoo. Now the sedge-warbler would not have been sitting unless it had laid a full clutch of its own eggs; all these had disappeared, and the presumption is that the cuckoo had removed them. Every cuckoo's egg laid close to the cottage in any year, and possibly in several successive years, would be those of one and the same bird. Yet this is the only instance in which I found a cuckoo's egg alone in a nest: and why she should have departed, on this occasion, from her usual practice of removing no egg, or only one from a nest, is not apparent. The sedge-warbler, it may be added, hatched the egg and reared the cuckoo.

The precocious prowess of the infant cuckoo in ejecting its companions and the method by which it does this are now matter of common knowledge, admitted and established as facts, and even photographed. Jenner described the process accurately towards the end of the eighteenth century, and from that time it was open to anyone who doubted Jenner's account to test it by observation or experiment. Yet more than a hundred years after Jenner's statement was made I read a signed letter in the press denying that an infant cuckoo ever did, or could act so. This denial is a good example of that failing in human nature that leads people to prefer to deny rather than to know. This disposition is often met with, but its opposite, the inclination to believe, or to assert without proof, is still more common. Both these tendencies are obstructions to scientific progress and to the spread of truth.

FAMILY LIFE OF THE MOORHEN

One habit which I have noticed only with the moorhen, or merehen, as it should more appropriately be called, may be worth recording. Perhaps it is well known, but I have seen it only in the moorhen, and have not come across it in accounts of other species; it shall therefore be given in some detail.

A pair of moorhens nest every year on a pond in the garden. One May, about the middle of the month, they had hatched their brood, and, as usual, picked up bread thrown to them and fed the young with it; so far there was nothing remarkable; but about the middle of July, when a second brood was hatched, the young ones of May still remained with the parents and assisted in the feeding of the July brood: they were now full-feathered and independent, but still young birds easily to be distinguished from the old ones. A parent bird would pick up bread, put it into the beak of one

of the tiny July young. The process was repeated again and again, and was a matter of amusement to us for several days. There were also diverting variations in the procedure. Once a May young one, having received a piece of bread from a parent, transferred it to the beak of another May bird, which then fed one of the little July birds with it. In this manner the

July brood were fed at second or third hand. Apparently it was against the rules for them to receive bread from a parent at first hand, for, when a parent bird did for once put bread straight into the beak of a July bird, one of the May birds at once ran up, took the bread out of the beak of the infant, and then replaced it there. 'Sheer red tape' was the comment of one to whom the story was related. The incidents described took place on the garden path within a few yards of my brother and myself, so that every detail could be seen. Another year a young moorhen of a first brood constantly took bread thrown to it and fed a young one of a second brood; in this instance the performance was in the water. In all these instances the moorhens were birds that had come to the pond by their own choice and were in a natural state.

It is said that unmated or bereaved old birds of various species sometimes join in the feeding of young birds that do not belong to them; but as far as my observations go, the example given by moorhens of this early development in young birds of a desire to undertake family duties is exceptional.

John Stuart Mill began at a very early age to teach a younger brother and sister, but he did not do this voluntarily, and he tells us that he greatly disliked doing it. The conduct of the young moorhens was voluntary and apparently agreeable to themselves.

CARE OF THE YOUNG: PARTRIDGES

As an example of parental care for the young by partridges, the following incident may be given:

A pair of old birds with their young were dusting in an open road on the Hampshire downs; as I approached on a bicycle they ran behind a straggling thicket of thorn and bramble. I dismounted, left the bicycle and went round the bushes after the covey to get a nearer sight of them. They were used to people passing along the road, and, not expecting to be followed, had stopped a few yards off on the edge of a large bramble patch. I came quietly and suddenly upon them; the young were in down, without a feather; flying was out of the question for them, and the covey disappeared at once into the impenetrable brambles — all but two of the little birds: these adopted the alternative concealment of squatting on the open grass. I stood still, with these two birds at my feet, and waited to see what the next move would be. There were slight movements in the brambles, and little sounds that suggested consultation between the old birds. The little birds outside remained motionless. After a few minutes one of the old birds, with great fuss and éclat, burst out of the brambles three or four yards in front of me, and invited pursuit by the usual manoeuvre of pretending to make frantic efforts to escape and to be unable to fly. My intention being to keep my eyes on the little birds at my feet, I did not stir limb or head; but so sudden and noisy was the old partridge's burst from the bush, that for one instant my gaze was diverted; in that one instant the two little birds at my feet disappeared: they were gone without my seeing them go: all was quiet in the brambles, nothing to be seen or heard of the covey except the single parent trying to decoy me away. Their manoeuvres were crowned with succcess by my withdrawing.

CHAPTER XI

WATERFOWL I

The grounds at Fallodon are not large. There is no park. There is no lake. There are two ponds, the larger of them less than an acre, a flower garden of fair size, and I have enclosed round the ponds two or three acres of rough ground planted with trees and shrubs. That is the place in which the waterfowl have been kept. Three things are necessary if you wish to have a collection of waterfowl. One is a fence as nearly fox-proof as you can make it, for the ingenuity of the fox is apt to defeat the very best and cleverest of human contrivances. In the enclosure you must have quiet, because waterfowl spend, in the early spring when they are in pairs, some weeks looking about for nesting-places, cautiously and quietly by themselves, and if they find that they are watched, or should you come suddenly upon them, and they are disturbed, they will not select that nesting-place, and will

not nest at all. So even in the case of one's self or the gardener, care must be taken not to walk at random in the nesting season on ground where birds are likely to nest, for fear of destroying the chance of their nesting altogether. Quiet is, therefore, the second necessity. The third thing necessary is that there must be some one who gives daily attention to the birds and takes an interest in them, and at Fallodon that has been done all these years by my gardener, Mr Henderson, to whose interest in the birds, and the great care he has taken of them, is due the credit of such success as has been attained in rearing the different species.

Now I come, in the first place, to the list of different kinds of waterfowl actually reared in the collection at Fallodon. Of British surface-feeding ducks, the mallard, wigeon, pintail, shoveller, garganey, and teal — in all six kinds of British surface-feeding ducks — have nested and reared young ones. Of British diving ducks, the tufted duck, red-headed or common pochard, the red-crested and white-eyed pochards — a total of ten kinds of British ducks. Of foreign ducks, the Spotted-bill, Carolina or North American wood duck, the Mandarin, Chiloe wigeon, Chilian pintail, Bahama pintail, Chilian teal, Rosy bill, Jalcated duck, Brazilian teal, blue-winged teal, Japanese teal, and versicolour teal — in all thirteen kinds of foreign and ten of British ducks which have, at different times, been successfully reared.

I have distinguished, in giving the list, between surface-feeding and diving ducks, and this, of course, is a very usual distinction given in books. But for the purposes of observation of waterfowl, one most remarkable distinction is between those drakes which have an 'eclipse' and those which have no eclipse. No doubt those amongst us who are interested in birds know what an 'eclipse' is. The most striking instances of the eclipse are to be found in the most brilliant-plumaged drakes. Take, for instance, the Mandarin drake, or the Carolina drake, which are two of the most brilliant-plumaged waterfowl in existence, in fact, two of the most brilliant of all birds in existence. The females of these species are quite sober and full-coloured birds, so that any one in the breeding season who knew nothing about birds, seeing the duck and the drake together, would hardly believe they were related to each other, so different are they in appearance as far as colour is concerned. But somewhere between the middle of May and the early part of June these brilliant drakes lose all their brilliant colours and become quite dull like the females, so that anybody who did not know the birds well would have very great difficulty in seeing the difference between the ducks and the drakes. It is a most remarkable and striking change.

*

Thus you have waterfowl divided, even in closely allied species — those which have an eclipse, and those which have no eclipse. And there is this curious difference which comes with the eclipse. Where the drake has an eclipse, he pays no attention to the female when the brood is hatched. For instance, the pintail is an early breeding bird, and sometimes it brings its brood on to the water while the drake is still in his breeding plumage; but whether he has come into the moult or eclipse or not, he never assists in the rearing of the brood at all. The Chilian pintail drake, on the contrary, which has no eclipse, goes with the duck and brood and takes just as much care of them as the cock partridge gives to a covey of partridges. I asked a friend, who had gone to South America to study wildfowl, if the same thing had happened in the wild state, and he informed me that as far as he had observed it was so. When the drake has no eclipse, he attends to and helps to bring up the brood. That is the curious difference of habit where drakes have no change of coloration during the year. What is the reason for it? Why should this difference result in the drake in one case helping to bring up the young, and in the other case paying no attention to them? These are matters which require much more study and which should be of much interest to those who have the opportunity of following them up. That is one point which is comparatively little noticed, so far as I have seen, in books, and it is of considerable importance to those interested in the habits of birds.

As to the actual breeding of the different species, I can give you in the time at our disposal only one or two instances of special interest.

I would tell you of one incident in the breeding of the Carolina or North American wood duck which I thought of considerable interest. I had a good many of these birds at one time, unpinioned, and therefore at perfect liberty to choose a nesting-place. Their natural nesting place is apparently a hole in a tree. One of my ducks selected a hole in an old elm tree some three hundred yards from the water. There she nested every year and brought out her young. The hole in the tree was a considerable distance above the ground, and Mr

Henderson (I was away at the time) was very interested to know how the duck managed to get its young brought down to the ground. One year he noted the day she began to sit, and, as he knew the period of incubation, on the morning the duck was due to hatch the eggs he sat down a little distance away opposite the elm tree. Presently he saw the duck come to the mouth of the hole and fly down into the long grass underneath, where she began calling. Then he saw the little ducks come to the edge of the hole and fall, one at a time, except in one instance where two fell together. There were six of them, and he told me they fell like corks into the long grass. Afterward I had the height from the ground measured and the depth of the hole in the tree measured. The hole was two feet deep, two feet perpendicular from the nest to the mouth of the hole. The hole was twenty-one feet above the ground, so that the little ducks, newly hatched when the mother flew out of the hole, had first of all in the dark cavity of the tree to climb up two feet within the trunk, then come to the mouth of the hole and throw themselves down, and after having done that to go with their mother for three hundred yards through the long grass following her to the water. I think this is a striking incident. Think of the little

ducks left in the nest. Newly hatched out, they had no feeding to strengthen them after leaving the egg. That they came out of the egg with such vitality and vigour that they could accomplish a climb of two feet perpendicular, and after falling twenty-one feet they could thereafter go three hundred yards through long grass, is a great tribute to the energy of Nature.

You will observe that the mother duck made no attempt

to carry them down. I have read in books that the common wild duck occasionally nests at a considerable height from the ground. I have seen one nest about seven feet from the ground, and know that this is so; but when I see it stated that in such cases the mallard or common wild duck carries the young ducks down to the ground, I doubt it. I think if any duck is in the habit of carrying its young to the ground, the North American wood duck would so so, as its natural nesting-places are in holes in trees and not on the ground like our common wild duck. Since this North American wood duck made no attempt to carry its young down to the ground, I am doubtful if any waterfowl would make the attempt at all. I will not say it is impossible. I have seen the young cuckoo, naked, helpless, and unable to stand up in the nest, turn out of the nest a newly hatched young bird that I put in with it; and having seen this, nothing to me in Nature is incredible. So while, judging from the wood duck, I am now of the opinion that no waterfowl carries its young down from the nest, I will only maintain that opinion until some trustworthy observer assures me that he has seen it done.

The versicolour teal which were bred at Fallodon were, as far as I know, the only birds of this species that have been reared in this country. Of course, I cannot be sure. There may have been some instance I have not heard of. These bred once with me, and the sequel is curious. Eight were reared, so I had a little flock of ten beautiful versicolour teal. The sexes are so alike, as is the case with several other South American waterfowl, that young males and females are difficult to tell. Unfortunately, out of the eight that were reared, six turned out to be drakes and only two were ducks. However, that made three pairs of versicolour teal. One pair I exchanged with dealers for something else which was rare and which I wanted, then the old duck which had bred died, and the young pair left were in the following year killed by a fox which somehow got into the enclosure. I found myself left with five drakes. Then came the War. Of course, during the War I made no attempt to buy any birds or replace losses by purchase. Two drakes I sent to the Zoological Gardens. They had not the species at all and were glad to have them. I had then three drakes. I heard of one female of the species being in the collection at Kew. I thought it worth while to send one of my drakes to Kew to mate with the female which had no mate, so I did that. In the next air-raid a piece of our own shrapnel fell and killed the female at Kew. Soon after that food became impossible to get, and what remained of my versicolour drakes, in common with several other rare things, perished. That completed the episode. First of all, the interest and satisfaction of rearing birds never bred here before; then the apparent security of having ten birds and the thought that I was sure to have representatives of this species in my collection to the end of my life; and now not a single representative of the species left to me. That is the sort of thing that happens with the rare and much desired.

One other point, I have selected to tell you about, the breeding of waterfowl. Of course, as you all know, wild ducks are monogamous and not polygamous like pheasants. They have one wife, and theirs is a very highly developed domestic life with great evidence of affection. Where the drake has no eclipse, the pair never separate during the year. Where the drake has an eclipse, he separates when in eclipse, and when he comes into plumage again, early in autumn, which most of the waterfowl do, though it is so long before the breeding season, the duck and drake come together again and spend the whole of the autumn and winter displaying every sign of affection in each other's company. The greatest instance I have seen of this is one that I will tell you. It was a red-crested pochard — a species very rare in Britain in a wild state. One drake that I reared was never pinioned, so that he could fly. I had him for over ten years, and during all that time he had never been away once. He mated with a duck, a bird of his own species, but which had been pinioned and could not fly. He spent years with her, and had every appearance of being happy and contented. One day, early in the year, his mate was injured by some vermin and practically ripped open on one side. She sat on the bank for two days perfectly helpless, and there he sat by her. She was so much injured that I had her caught and put out of her pain. There was another female pinioned, red-crested pochard unmated, and I thought, of course, that the now lonely drake would mate with her; but he would pay no attention to her. He spent, if I recollect the time — it was some years ago — two or three weeks flying about with every sign of restlessness and distress from one pond to the other looking everywhere for his old mate. I had had him for some ten years, and he had never gone away, but now after two or three weeks he went. He flew away, and I never saw him again; it was as if he had gone on an endless search of the world for the mate he had lost. That sort of thing is very interesting, for it shows the great natural affection which exists amongst birds of a highly developed and intelligent species. To me it is a clear proof of the fact that the relationship between the more highly-developed birds is one of real domestic happiness, not confined to the breeding season and the reproduction of species only. I know that swans become attached to each other. You can see it is so. They do become permanently attached to each other, and have domestic happiness, which plays a large part in their lives, quite apart from the breeding season.

Perhaps you would like to know how long these sorts of birds will live. A great many of my birds are unpinioned and fly away, but in the case of a pinioned bird you can tell how long it lives. The longest-lived bird I had was a Chiloe wigeon drake. I bought him as a fully-grown bird in October, 1888, and he died peacefully, and obviously of sheer old age, in October, 1908. I do not know how old he was when I bought him, and this is the longest life I have known of any of my waterfowl. For some years before he died he had shown very distinct signs of old age. He became very stiff, and infirm, and eventually was found sitting on the bank dead, obviously of old age. Geese, and larger birds probably, live much longer.

Since I have had these waterfowl of different kinds at Fallodon, it has been very interesting to see the varieties of wild ones which have come to my ponds. I remember when I was a boy my father showing me a place on one of the burns at home, and saying, 'That is the place where I once shot a teal'; and with that and one other exception, nothing but mallards, as far as I know, have ever been shot or seen on the actual property at home. It does not extend to the sea, and the sea ducks do not come to it; but I myself once, after a great gale in the winter, shot an immature wigeon on a little pool. With these exceptions, nothing but the common wild duck used to be seen on the property at all. Now every year my ponds are visited frequently by the mallard, teal, wigeon, pintail, shoveller, pochard, and tufted duck. I treat the enclosure as a sanctuary. That shows how so many kinds considered rare by those who shoot, such as the shoveller, are often passing over, especially in the season of migration, and, if they hear birds of their own kind calling below, come down and settle. One very interesting point about wild things is how quickly you can get a perfectly wild bird tame. I remember one December afternoon finding a wild pintail drake on the pond. He rose, flew high into the air, and circled round; but when he saw that the pinioned and tame birds did not follow him, after much flying at a great height he lit again on the pond. That evening when I was feeding the birds, he came and looked on, and within a week he would come out with the others to feed and pick up the grain I threw to him, and even when some of the grain fell on his back he was not alarmed.

So you see how tameness in their own kind gives confidence to the wildest birds, but that tameness, that confidence, is associated with the place, and does not cause them to be less wild elsewhere than they were. I had one good instance of this in the case of a drake shoveller. A brilliant bird in full plumage, he came one year in February or March and stayed on my ponds. He was not always there, but he was often there, and he adopted all the habits of my tame shovellers. If he was sitting on the bank and I walked past, he would fly five or six yards into the water and sit quite unconcerned. If he was in the water, he did not offer to get on the wing at all. One day after lunch I walked round the pond and saw he was not there. I went for a bicycle ride, and coming back, about a mile from home, I saw on a pool in a field, not on my property, a shoveller drake in full plumage. I felt morally certain it was the same drake which had come to my ponds, because you do not commonly see a wild shoveller in the district. The pool was about one hundred yards from the road. I got into the field and walked straight towards him. He rose off the water, went high in the air, and after circling about I saw him go straight for my woods. I stepped the distance at which he had risen, and, allowing for a few yards of water, those I could not step, the distance was something over ninety yards. I bicycled home, and went straight to the farther pond which he usually frequented, and there he was on the water, perfectly unconcerned and tame. That is a very good instance of how quickly birds find out when a place is a sanctuary. Really it is worth while for anybody who has a little bit of ground and a little patch of water to have a few waterfowl and make it a sanctuary.

One more instance I will give of wild birds tamed like this. I came home once in December after an absence of some months and found that there was a pair of teal which had been in the habit of coming to feed with my waterfowl when no one was present. They were wild, and when I showed myself at one pond they flew to the other, and so forth. Gradually, however, they began to come out and watch the feeding of the others every evening. They came out of the pond about twenty yards from the spot where I was sitting.

They came cautiously round behind the trunk of a beechtree, by a route that none of the other birds took, and drew nearer and nearer every evening until, after some four weeks, they came up a little behind me, a little to the right and about four yards from me, and I got them to pick up the grains I threw. That went on for some weeks. In March they went away, presumably to breed somewhere else. In August I was standing at the end of one pond when a drake teal in eclipse came flying and lit about ten yards from me. I thought, of course, that he had not seen me, and as soon as he saw me he would fly off. He took a look round, and when he saw me he never offered to fly away, but began to preen his feathers. I thought, 'That is the little drake come back again. I shall know at feeding time in the evening'. At feeding-time he came out with the other birds and picked up the grain.

I could not be sure he was the same drake, but about the duck I can be sure. The drake stayed, came out of the eclipse, and got his full breeding plumage, but there was no sign of the duck until November. In November I was walking along the bank when a little duck teal flew off the bank and out on to the water a few yards. It remained there while I passed. I thought, 'That must be the little duck come back for the winter. I shall see her at feeding time in the evening'. In the evening I sat at the foot of a big larch as usual, and the little duck teal came out of the pond, not as the others came, but about twenty yards from the others, just in the same way as she had come in the previous winter. She came round the trunk of the beech by a route no other ducks took, and stood at the same angle and at the same distance as she had done before, and began picking up the grain I threw.

These little things — and I have had a great many instances of them — if you have an interest in waterfowl, these little things add extraordinarily to the interest of life. The sequel is a short one. She paired at once with the little drake, and I hoped they would stay through the winter. After three weeks they flew away. It is some years ago now, but for some time I cherished the hope that they would come back again.

My collection now is not what it was. I lost a great many birds during the War owing to bad food. I did not attempt to replace losses, and could not have done so in some cases had I tried. I have got a lot of birds still, mostly bred on the place, but I have not the same number of species that I had; and now, partly because my sight is so much impaired, I find most interest in having as many as I can unpinioned, as tame as possible, and yet at perfect liberty. There is a sort of romance in having naturally shy birds perfectly free and unpinioned, coming, as some of my wigeon and pintail do, to feed with perfect confidence out of my hand, while I know all the time that any day they may join the wild ones to go south in the winter or far north in the spring. They are naturally shy things, yet, when they are in this particular sanctuary, they are perfectly tame and have perfect confidence. If they should go away with the wild birds, they will be wild outside like others of their own species, and yet any morning I may go round the pond and find they are back quite tame again. There is interest, almost romance, in these happenings, and I know of no greater satisfaction for people interested in birds or animals than having wild things altogether free from control, naturally wild and shy, yet perfectly tame, so that they show you confidence and trust. You can observe their natural habits going on in your presence, which you could not do with the same species in the wild state, because their exceeding shyness and fear of man make it impossible freely to observe their habits till in some way or another you have convinced them that in one place at any rate man is not an enemy. Then you can have the great pleasure of watching close to you the colour of their plumage, their movements, their courting, their flight, and all the things that make them beautiful and interesting.

CHAPTER XII

WATERFOWL II

IN 1921 I gave some account to a Naturalist Society of the collection of waterfowl at Fallodon. [This is reprinted in Chapter XI.]

*

It is not intended to repeat here any of the substance of that address; but more than five years have passed since it was delivered, and it may be of interest to give some account of incidents that have happened and observations that have been made in these later years.

It has been said that in species in which the drake has no eclipse and the plumage of the two sexes is alike, the drake assists in rearing the young brood. Further experience has confirmed this. There was a remarkable instance at Fallodon in 1926. A Chilian teal nested some fifty yards away from the water in the broad box edging of a flowerbed. The nest was close to a much-used garden path, but it was completely concealed by the box-leaves, which met above it. The bird entered through a hole at the bottom of the clipped box edging on the inner side. Seven eggs were laid, and were hatched in May; both duck and drake were always with the brood, the drake as inseparable from and as devoted to the young birds as was the duck. Four young ones were reared. I succeeded, with the help of the parents, who were pinioned birds and very tame, in training them to feed from the hand, but they were left free, and no wing was clipped or pinioned. When the young birds were well on the wing in July, the duck was found to be laying again in the same nest; she stripped a fresh lot of down, the clutch was again seven eggs, and she hatched them all, on the 5th of August. The drake was as attentive as before, this time six birds were reared; and not till they were well on the wing did the family break up.

It is, I imagine, rare for any wild duck to rear two broods in one season; and it could not happen in these species in which the drakes have an eclipse. Indeed, this incident is a remarkable example of the continued vigour, as well as the family affection of the drake Chilian teal. Any drake that had an eclipse would, by the end of May or early in June, have become dowdy, depressed and indifferent to his mate. The Chilian teal, on the contrary, remained as sprightly and gallant through the whole summer as he had been in the spring.

*

When my first waterfowl were brought in March 1884, it seemed to me no drawback that the birds were pinioned. My interest, like that of most collectors of waterfowl, was centred on variety of species, their plumage and their ways. Then came a wish to get the birds to nest, and great care was taken to rear the young: this was often done by taking the eggs away from the duck's nest and setting them under a bantam for greater security both in the nest and in subsequent protection of the young birds. In later years my point of view has changed; pinioned birds are no longer a complete satisfaction. It is not on humanitarian grounds that discontent has arisen: from all that I have seen and been told, it is evident that the pinioning of very young ducks does not cause suffering; they feed and show no sign of pain or discomfort after the operation, nor do they suffer appreciably in later life. Their plumage, health and spirits are good: it is no uncommon event for an unpinioned bird to pair year after year with the same pinioned mate; the two birds lead the same life together, and it cannot be much hardship to the pinioned bird to be compelled to remain in an enclosure in which the free bird stays voluntarily.

Nevertheless, pinioning is a blemish to the appearance, and more and more have my feeling and sentiment inclined to free birds. The pinioned bird is a prisoner: one cannot look at it and reflect that it stays by its own choice, that it remains by preference and not necessity. A guest once remarked to a great landowner on the promptness with which curtsies were dropped or hats touched by every one they passed. 'Yes,' said the great man, 'they know very well that I *will* be popular on my own estate.' With pinioned birds there is a doubt as to the reason for their complacence: with free birds their tameness is a compliment.

I have therefore in recent years very greatly increased the proportion of free to pinioned birds at Fallodon, and by regular and patient attention on my part or on that of others when I am away from home, the standard of tameness has been raised to a very high plane. There may now be seen on the wing at Fallodon pintail, wigeon, tufted ducks, pochards, red-crested pochards, shovellers, Chilian pintail,

Chilian teal, Chiloe wigeon, Mandarins, and wood ducks *(Carolinus)*. Of these free birds a considerable number, representing every one of the species named, except the shovellers, will come out of the water and take food confidently from the hand. In order to get this result many hours are spent, particularly in June, in taming brood after brood of young ducks: for it is when they are quite small that the training must be done. Even when the parent birds are tame, it often needs many days before their example and reassuring notes can overcome the innate suspicion that the little birds feel of a human hand.

The disadvantages of having rare waterfowl on these terms are serious. Numbers of them fly away and never return; many, no doubt, are shot; for no one, however well intentioned, can distinguish a rare duck when flight-shooting.

The number of birds that disappear every season gives an uneasy feeling of spendthrift waste; for £5 or £6 is not an unusual price for a pair of the rarer species. The mere absence of the birds would not matter: the thought which hurts is that so many of them leave, not to be beautiful and interesting elsewhere, but to be shot unrecognised. It is seldom that I know definitely of the catastrophes, for none are ever shot on my own property. Two young Chiloe wigeon, both hand-tame, left me one October: they never returned, and in the following January the wing of a rare duck was sent to me for identification. It was the wing of a Chiloe wigeon in the plumage of its first year. The bird had been shot about twenty miles north of Fallodon. Chiloe wigeon in a free state are, I imagine, uncommon in Northumberland, and probably anywhere else in England, and there is little doubt that this wing was that of one of my young birds. Two points of interest attach to this occurrence. One is that the bird had not migrated and gone South, though it left me when the Southward tide of migration was strong. As it is a South American species, its migratory instinct may well have been confused in the Northern Hemisphere. The second point is, that even in winter these young birds of a foreign species had been able to get food so easily that they never found it worth while to return to Fallodon to be fed, though they had remained within easy reach of it.

Another drawback incident to the freedom of the birds is that they settle for themselves the numbers of each species that are to remain with me; and as my accommodation in water is small, there are too many birds of certain species. Tufted ducks are prone to what Peacock calls 'stay-at-home-ativeness'. The young broods are apt to be content with the place in which they have been reared. The result is that day after day I have to look at a flotilla of twenty or thirty tufted ducks, sitting content, comfortable and secure on the limited space of water which all the other waterfowl have to share. Most of the tufted ducks could well be spared and would be better bestowed elsewhere. Yet to interfere is impossible. Though they are all free birds, they could be caught by feeding them into a covered wire enclosure; but this catching would shock the whole collection, destroy their confidence, and undo in one fell moment the work of years spent in inducing the birds to believe that the place is safe and that mankind in this one spot can be trusted.

Some eight or ten of these tufted ducks will stand at my feet looking up in the most engaging way to be fed by hand: in the confusion of the catching, these, from which I should not wish to part, could not be distinguished from others that I should not miss. So the tufted ducks increase out of due proportion to the rest, and there is nothing to be done but to acquiesce. They must be well fed, too, for to stint them would be to stint other kinds, which are more rare and few in number and which it is desired to keep. One becomes positively servile in pampering free birds that one fears to lose.

In contrast with tufted ducks are pochards: these are very considerate and never take too great advantage of hospitality. Young pochards leave before the end of October; it is an exception when one of the young birds stays with me for the winter. Tameness makes no difference; however intimately tame a brood may be, the instinct to depart is the dominating influence. Two pairs of free pochards have bred at Fallodon every season in the last two or three years. Of these, I think the females do not go away at all. The males leave when the females are sitting in May; this year (1926) one of them returned at the beginning of August, and the other about three weeks later. The first one could be identified with certainty, for he fed from the hand and had precisely the same manners as before he left. The second bird was not hand-tame, but would feed within a yard of me, and was evidently used to all the ways of the place: I assume, therefore, that he was one of the birds that had been here in the nesting season.

Pochards and tufted ducks do not seem to mate till after, and generally long after, the new year, whereas many wigeon, pintail, Mandarins and wood-ducks come together soon after the eclipse is over, and court and go in pairs during the autumn; and thus have a longer period of domestic happiness. It is remarkable that male and female pochards should be so late in taking notice of each other, for the drakes have only a partial eclipse, much less marked than that of the other species with which they are here compared.

Every year I suffer the disadvantage and losses of having so many unpinioned waterfowl,but more and more the pleasure of having free birds tame preponderates.

To see a pair of Chiloe wigeon on the wing high in the air, conversing audibly as they fly, and to know that when they alight they will take food from the hand — all this is a satisfaction that no pinioned birds can give. To have even one such pair that stays is compensation for the loss of several that fly away and do not return.

In the last few years there are been many birds that by some individuality or ways of their own have become subjects of special interest. To each of these a little history belongs. One shall be selected as an example, both of the pleasure and the pain of having tame birds to which one becomes attached. I will call it the story of Elizabeth.

Elizabeth was a pintail; one of a brood reared in 1921 by an unpinioned mother that had paid the enclosure at Falloden the compliment of nesting there. Several of the young pintails reared this year would take bits of bread from the fingers, but only three, one male and two females, would put their bills into the palm of the hand and thus eat their grain. These three birds continued to do this through the autumn and winter. In the spring of 1922 I lost track of them, and the male bird has not returned, or at any rate has not been identified again. But in the early autumn of 1922 the two females began to attend the evening feed regularly, and ate grain from the hand as before. One of them was rather darker in plumage than the other; in the spring of 1923 both disappeared again, and after the nesting season was over only one, the darker bird, presented itself and fed as usual. There were several other pintails, and they were all tame, but this was now the only one that ate corn from the hand: it became convenient now to have a name by which to speak of it, and thus it came to be called Elizabeth.

Elizabeth, when so disposed, would take food from the hand at any time and in any part of the grounds; but her favourite and regular place was to the right of the low bench from which the evening feed was given at the foot of a big larch. A root of the tree heaves up the ground almost level with this end of the bench, and here she would stand. If in the business of feeding the other birds her arrival was not noticed, she would make a little noise that sounded like a gentle reproach: on turning my head I would see her standing there, looking up at me expectant, till she was satisfied by a handful of grain held out to her. Every evening her presence was looked forward to, and if, as sometimes happened, she was absent for a day or two, there was a sense of disappointment and of something missing.

In spring the waterfowl do not attend the evening feed regularly: they are scattered and busy with nests; but in the spring of 1924 Elizabeth still came now and then, and it was evident from her manners that she had a nest and was sitting. A nest under a big thorn bush in a field some three hundred yards away was thought to be hers; but as there were other pintails sitting in the neighbourhood, there could be no certainty whose nest it was. It was robbed of all the eggs by vermin. Elizabeth never brought off a brood, and about the end of May she left. Not till November was she seen again, and then one evening there was the little note of reproach, and Elizabeth was standing in the usual place beside me. The unexpected return of a favourite bird after months of absence, during which it must have led a perfectly wild life and perhaps visited the Arctic zone, uplifts one with satisfaction. This year she remained as usual throughout the winter; but in the spring of 1925 she disappeared, presumably to nest elsewhere. Each season she had a mate, who would stand close by, but would not feed from the hand with her: whether it was the same drake every year was uncertain. This year she came back very early in the autumn, but in December she went away. It is a dangerous time, for there is much duck-shooting. One evening in January she reappeared at the evening feed in very sad case. She must have been able to fly, or she could not have come back over the fox-proof fence, and neither leg was broken; but both wings dragged, and she had a crouching walk. She put her bill in my hand, but seemed unable to swallow the corn in it and tried ineffectually to eat the grains that were scattered on the ground. The next evening she came once more, and this time I had some soaked bread ready for her. I gave her this by hand, and she ate a fair amount; but she was not better, and was never seen again.

It is impossible to have tame birds without facing these tragedies: one has to make up one's mind to do so, and to reflect that over a long series of years the tragedies are intermittent, whereas the interest given by the birds is a daily pleasure which is not only unflagging, but increases. Nevertheless the tragedy when it comes is very poignant. As to Elizabeth herself, the injury, whatever it was that befell her, must have been one that might happen to any pintail in a natural state; she was in her fifth year; that may be less than the average length of the life of a wild pintail; but for all the months of it spent at Fallodon, she had been more secure in severe weather from hardship such as scarcity of food, and less disturbed by fear and alarm, than wild birds of her kind: and for the rest, she had by going away at will been able to satisfy all the instincts for wild life.

Year after year there are like incidents of individual interest, of which the story of Elizabeth is but one example.

*

Some part of the pleasure given by tame waterfowl is due to the reflection that many of them are by nature so exceedingly wild and distrustful of man. Mr J. G. Millais, in his valuable and magnificent volumes on British ducks, notes the very great difficulty of taming pintail and teal, and my experience agrees with his, except that I should say that teal were the more difficult. Only twice in forty-three years have pinioned teal nested at Fallodon.

Of pintails Mr Millais says: 'Even in confinement ... if whoever feeds them ceases to go near the pond for a few days, they seem to resume at once the instinctive fear of the wild state.' He says also: 'Pintails will never approach closely unless men are constantly around them, as in such public places as St James's Park and the Zoo.'

This is the character of birds caught wild and pinioned; but it is possible to rear a race of pintails that remain tame. Every autumn from six to a dozen come to the two ponds and remain till the spring. There are at least three pairs here now (January 1927). It happens that in 1925 and 1926 only one pair of pintails stayed to nest at Fallodon, and in each year the duck brought out her brood early in May in very inclement weather, and they all perished. No pintails have been reared at Fallodon since 1924, and the youngest birds among those now here must be at least in their third year. Every one of them has, for several months of each year, been absent, and led a perfectly wild life. When away they are probably as shy and distrustful of man as wild birds of their kind, but when again on the Fallodon ponds they are as tame as ever. Only one of those now wintering there will feed out of the hand, but they will all come to within six feet of me on land, and there feed and court and bicker with each other in a perfectly natural way. Long absences and wild life have not destroyed their confidence in a place that they have been brought up to believe to be safe. This instance of the pintails has been thought worth mentioning because of the reputation of the species for wildness. London provides examples on a large scale of waterfowl and gulls becoming tame in a place where they know that they are safe, but I have not myself seen pintails there behaving like tame birds; though others with better eyes and opportunities may have done so.

Tame as the Fallodon waterfowl have become — some of them will fly up and stand on the seat beside me to be fed — the wild instinct is always there. Any unusual movement upsets them: it does not do to knock out the ashes of a pipe against the seat when they are round it; and to put up a parasol in their presence causes disaster; on the other hand, one may put up a pair of binoculars without disturbing them, for they are used to this. The thumb may stroke the breast of a bird that is feeding out of the hand, but any attempt to stroke the back might never be forgiven. Their resentment of any unusual familiarity reminds me of the following story which my grandfather used to tell of Sydney Smith. In some party where he was present, one of the company had received and accepted an invitation to stay at Lambeth Palace. This man was somewhat forward in manner, and was talking with assurance and familiarity of his impending visit, though he had never been at Lambeth before. As they parted, Sydney Smith said to him, 'Let me give you one little bit of advice: when you get to Lambeth don't clap the Archbishop on the back and call him "Canterbury", because he isn't *used* to it.'

One disadvantage of having pinioned birds is that hybrids will sometimes occur. One cannot always ensure that every pinioned bird has a mate; pairs are broken by one bird dying or meeting with mishap, and with rare kinds the loss cannot always be replaced at once. This leads to *mésalliances.* They are not very common, and when they occur Nature takes care that the offence is not perpetuated. I have noticed that hybrid drakes, the offspring of two kinds not closely related, have no clear voice, nothing but a sort of husky whisper.

Closely related birds of the mallard species will produce any number of fertile and ugly mongrels. I have, however, had experience of one hybrid among other kinds which has been sufficiently interesting to merit some attention.

In 1917 a male pinioned white-eyed pochard paired with a pinioned mate of his own species, and, as far as could be observed, was faithful to her in the early part of the season: but when she had been sitting on eggs for some time, he courted an unpinioned tufted duck; he swam with this bird, sat by her on the bank, and consorted with her all day; she in time nested, and reared two birds of her brood. They were left free, and proved to be a drake and a duck; they paired and went away together for a while in the autumn; but returned, and though occasionally absent, made the enclosure their home. They did not breed in 1918, but the female nested in 1919, and that year and every year since she has nested and brought a brood to the water; the younger generations have also been fertile, for though most of them have gone permanently into wild life, there were at least three nests of these hybrids last year (1926) at Fallodon. The broods are as numerous and as vigorous as tufted ducks. These hybrids have so far paired entirely with each other, and they breed true to type. Each generation resembles the first pair. The original male has had the pecularity that he will not touch bread: he will look on while the other birds eat it; but however close to him a piece falls in the water he will not take the trouble to pick it up. No other of my tame waterfowl is so uniformly disdainful, and it is by this habit that I have been able to identify him. The female was not an exception in this way, and it is not therefore certain that it is the original female with which the drake still pairs; but I have assumed that this is probable, and that she is one of the hybrids that have nested every year. Both sexes show very conspicuously the white under-parts at the end of the body that is characteristic of white-eyed pochards; the females have the dark eye of a female white-eyed pochard, but the plumage is not so rufous and they are larger birds. The males have eyes distinctly paler than those of a tufted drake, and they have no tuft on the head. On each side, when in plumage, they have a panel corresponding to the white panels of tufted drakes; but in the hybrids this is a soft grey. The black on the head in sunlight has a green instead of the violet sheen of a tufted drake. The drake hybrids have a complete eclipse. The notes, so far as I have been able to observe, are not like those of tufted ducks, but very closely resemble those of white-eyed pochards. So far these hybrids exhibit the characteristics of a separate and permanent species.

There are other pleasures in watching waterfowl besides that of attention to individual birds or species. Wordsworth in 'To a Highland Girl' notes the precious power of memory to preserve a vision of beauty or the impression of the mood that accompanied it. I will try to describe one such scene, as memory recalls it.

It was Christmas morning, many years ago when my sight was less impaired; the sun does not rise till after half-past eight in Northumberland at this season. I went out after breakfast; the waterfowl had been fed rather later than usual:

they were still finishing their meal under the big larch tree or were assembled at this end of the pond, which, being much shut in by trees and shrubs, was still in dark shadow. I went to the other pond some two hundred yards away and sat on the garden seat on the farther side. This pond is more open: there are no tall trees on the east side, and all the water was in full sunlight. There was not a bird on it; there was no stir in the air; the surface of the water was smooth and without motion. Presently pintail, wigeon, tufted ducks, pochards and one or two other kinds began to come flying over the intervening shrubs and threes from the pond where they had been fed. They came, some singly, some two or three together. None of them had yet seen the sun that morning, and each and all, as if in greeting to it, began to sport and play. They threw the water over their bodies, they raised themselves up on it and flapped their wings: they swam rapidly about in all directions, low in the water with quick and eager forward dartings of the head and neck. They sprang from the water into the air and took headers from the air into the water; they made short flights in one direction, lit on the water for a moment, and made another flight back in the direction whence they came; they dived unexpectedly, travelled under water, came up in some new place, and then, as if surprised at what they saw, dived again with exceeding suddenness. They splashed for the sake of splashing; there was not a square foot of water that was not in constant agitation. For some time the scene was one of motions of delight and exhilaration. At length first one bird and then another flew up on to the bank that faced the sun, or on to the south end of a little island; there they stood or sat, many of them side by side in pairs, and rested motionless or slept. Some half-dozen birds only remained on the water, and each of these was still, the head turned round and the bill resting in the feathers of the back. All was quiet; there was no sound or stir; the water was again smooth, the reflections in it were composed once more; the sun still shone; on the water and the birds; on the scarlet-barked willows and the delicate bareness of winter trees on the opposite side. Any one who had come upon it now might have thought that the place was under some spell. He would have seen the man on the seat motionless, too, for a long time; entranced rather than asleep: the scene had indeed sunk down into his heart and 'held it like a dream'. There are times when man's consciousness seems laid to rest in some great whole, of which he has become a part. There are hours of which it can be said, 'Thought was not: in enjoyment it expired.' So it was not, and if anything stirred in the mind at all, it was an echo of the words, 'And God saw that it was good.'

SOURCES

The extracts in Part 2 are taken from the following works by Grey:
Fallodon Papers, Constable & Co, London, (1926)
The Charm of Birds, Hodder & Stoughton, London (1927)

The material has been extracted as follows:

Chapter V	From pages 69-79 of the *Fallodon Papers*
Chapter VI	From Chapter 1 of *The Charm of Birds*
Chapter VII	From Chapter 2 of *The Charm of Birds*
Chapter VIII	From Chapter 3 of *The Charm of Birds*
Chapter IX	From Chapters 3, 5, 7, 9, 10 & 11 of *The Charm of Birds*
Chapter X	From pages 34-37 of the *Fallodon Papers* and Chapters 8, 9 and 11 of *The Charm of Birds*
Chapter XI	From Chapter 5 of the *Fallodon Papers*
Chapter XII	From Chapters 13 & 14 of *The Charm of Birds*

ACKNOWLEDGEMENTS

Two distinguished professors of modern history have written biographies of Grey: the late G.M. Trevelyan in 1937 and Keith Robbins in 1971. Their influence will be readily apparent in the pages which follow and is gratefully acknowledged.

Many thanks are due to Mr Peter Bridgeman, the current owner of Fallodon Hall. The description in Chapter IV owes much to his courtesy and kindness.

Thanks are also due to Miss Henrietta Ryott of the National Trust and Mr Ian Dawson of the Royal Society for the Protection of Birds.

Jan Karpinski

All illustrations in the book, with the exception of pages 8 and 45, are by Robert Gibbings and originally appeared in the following publications:

'Where I Lived and What I Lived For' by Henry Thoreau, Golden Cockerell Press, 1924; 'Fallodon Papers' by Viscount Grey of Fallodon, Constable & Co. Ltd., 1926; 'The Charm of Birds' by Viscount Grey of Fallodon, Hodder & Stoughton, 1927; 'The Hundredth Story' by A.E. Coppard, Golden Cockerell Press, 1931; 'The Roving Angler' by H.E. Palmer, J.M. Dent & Sons Ltd., 1933; 'Lord Adrian' by Lord Dunsany, Golden Cockerell Press, 1933; 'Beasts and Saints' by Helen Waddell, Constable & Co. Ltd., 1934; 'A Bird Diary' by Godfrey P. Harrison, J.M. Dent & Sons Ltd., 1936; 'The Twelve Months' by Llewelyn Powys, John Lane The Bodley Head Ltd., 1937; 'A Book of Uncommon Prayer' by George Scott-Moncreiff, Methuen & Co. Ltd., 1937; 'The Art of Living' by André Maurois, The English Universities Press Ltd., 1940; 'Sweet Thames Run Softly' by Robert Gibbings, J.M. Dent & Sons Ltd., 1940; 'Coming Down the Wye' by Robert Gibbings, J.M. Dent & Sons Ltd., 1942; 'Coming Down the Seine' by Robert Gibbings, J.M. Dent and Sons Ltd., 1953; 'Till I End My Song' by Robert Gibbings, J.M. Dent & Sons Ltd., 1957; 'The Wood Engravings of Robert Gibbings' edited by Patience Empson, J.M. Dent & Sons Ltd., 1959.

We are grateful to the publishers of these books and the Robert Gibbings Estate for permission to reproduce the illustrations. We should, also, like to thank Patience Empson for all her help and for making many of these illustrations available to us.

The illustration on page 45 is reproduced courtesy of Peter Bridgeman and the photograph on page 8 courtesy of BBC Hulton Picture Library.

INDEX